5.20.77

The Complete Handbook of
Slow-Scan TV

No. 859
$14.95

The Complete Handbook of
Slow-Scan TV
By Dave Ingram

TAB BOOKS
Blue Ridge Summit, Pa. 17214

FIRST EDITION

FIRST PRINTING— APRIL 1977

Copyright © 1977 by TAB BOOKS

Printed in the United States
of America

Reproduction or publication of the content in any manner, without express permission of the publisher, is prohibited. No liability is assumed with respect to the use of the information herein.

Library of Congress Cataloging in Publication Data

Ingram, Dave
 The complete handbook of slow-scan tv.

 Bibliography: p.
 Includes index.
 1. Slow-scan tv.   I. Title.
TK9960.I53.   621.388'5   77-3735

ISBN 0-8306-7859-X
ISBN 0-8306-6859-4 pbk.

Preface

A new sound is being heard more and more often on the high frequency amateur bands—slow-scan television. Behind this sound lies a complete new era of visual communication which presents unlimited capabilities for the amateur radio operator.

Although the open-air days of convertible automobiles, biplanes, and breadboard rigs are slowly becoming a memory, the golden age of radio still exists in the world of SSTV. Imagine the excitement as you, possibly the only radio amateur in your area with video capability, and your friends watch pictures transmitted directly from various areas of the world. Visualize acquiring your first look at someone with whom you've talked many times, or showing your rig and special projects to another SSTV operator who is genuinely interested in your work. SSTV is all the excitement of your first QSO, most outstanding QSO, longest DX QSO, most newsworthy, and record-breaking QSO all rolled into one. Truly this is ham radio supreme!

Now is an ideal time for the radio amateur to get involved with slow-scan TV. Present interest in VHF bands, coupled with adverse sunspot activity, have left the 80- through 10-meter bands wide open to worldwide communications. DX contacts are not suffering from heavy QRM and clear frequencies for SSTV can be located with little trouble. As SSTV moves into the limelight on these bands, a respect for modern pioneering technology is becoming apparent.

In an attempt to provide an absolute state-of-the-art analysis, I set about writing this book while leading SSTV designers around the country perfected their particular units. Our efforts merged at publication time when this book took final form. Until this time, the book was somewhat of a dream.

I have prepared information in this book following a logical sequence. The first four chapters discuss SSTV in an introductory form. Simplicity is the keynote to these chapters as the newcomer learns of SSTV systems and parameters. The latter part of the book reveals many revolutionary new ideas and circuits which reflect the latest SSTV innovations. While many of these units may appear somewhat complicated, constructional obstacles are overcome through the use of printed circuit (PC) boards of each unit. Thus in most cases, the particular designer of a unit may be contacted directly for PC boards of his unit. Exceptions to this system are noted, where certain other individuals handle PC work for designers.

Creating a book of this type is not a simple matter, and I would like to express sincere gratitude to the following individuals for their support and contributions to this book: Dr. Don Miller, W9NTP, of Waldron, Indiana, Dr. Robert Suding, W0LMD, of Lakewood, Colorado, Mike Tallent, W6MXV, of San Jose, California, Bob Tschannen, W9LUO, of Lombard, Illinois, and Bob Schloeman, WA7MOV, of Phoenix, Arizona. Overseas contributors included: Professor Franco Fanti, I1LCF, Joseph Trombino, CP1BCC, Barry Sharpe, VK5BS, and Franta Smola, 0K100. Many of the off-the-monitor photos contained in Chapters 1 through 5 were shot by Beverly Taylor, a superb local photographer. My special thanks to this busy person.

Thanks also to the following SSTV manufacturers for permission to use information from their manuals: Robot Research Inc. of San Diego, California, Venus Scientific Inc. of Farmingdale, New York, Sumner Electronics and Engineering of Hendersonville, Tennessee, and Linear Systems Inc. of Watsonville, California.

Finally, my greatest appreciation goes to Sandy, WB4OEE, who was such an outstanding inspiration to me to write and complete this book.

Dave Ingram, K4TWJ
Birmingham, Alabama

Contents

1 Introduction To SSTV — 9
SSTV Around the World—Summary

2 Understanding SSTV Gear — 30
SSTV Monitors—Home Brewing SSTV Monitors—Flying-Spot Scanners—SSTV Camera—Sampling Cameras—Direct Scan Conversion—Digital Fast- to Slow-Scan Conversion—Digital Slow- to Fast-Scan Conversion

3 Setting Up the SSTV Station — 48
Monitors—Cameras—Tape Recorders—Taping Procedures—The SSTV Studio

4 Operating Procedures — 65
Producing Programs—SSTV DX—Power Requirements—Live Transmissions—Framing SSTV Pictures—Operating SSTV Without a Monitor—SSTV Picture Analysis

5 SSTV Monitor Circuits — 81
W9LUO Mark II Monitor—RF Feedback

6 Digital Scan Converters — 106
Elementary Principles of Digital TV—W0LMD Digital Scan Converter—MXV-200 Scanverter—D-A Converter—MXW-22 Parts List—W9NTP Slow-Scan Color Converter

7 Roundup of Existing Gear 190
Buying Considerations—Robot 70 SSTV Monitor—Robot 80A SSTV Camera—Robot 60 and 61 Viewfinders—Robot 300 SSTV Scan Converter—Robot 400 Converter—Venus SS-2 Monitor—Venus C1 Camera—SEEC HCV-2A Monitor—SEEC HCV-1B Camera—SEEC HCV-3KB SSTV Keyboard—SBE SB-1 MTV SSTV Monitor—SBE SB-1 CTV SSTV Camera

8 SSTV Satellite Communications 270
Weather Satellite Pictures using an SSTV Monitor—Weather Satellite Receiving Station

Glossary 299

Index 302

Foldout Section 193

Chapter 1

Introduction To SSTV

There is a tremendous fascination in actually watching pictures, transmitted from places thousands of miles away slowly roll down an SSTV monitor screen. An amateur radio station equipped with SSTV becomes a personal travel bureau right in your living room or den. Not only does the addition of video add warmth to a QSO, it also serves to expand knowledge on a worldwide basis. You, as an SSTV operator, will soon find this visual capability allows a self-expression previously impossible. An automobile buff transmitting SSTV pictures of vintage cars might find himself viewing friends' information and pictures that many magazines would relish. Through the medium of SSTV, an Australian farmer might enjoy his first view of an African sunset (Fig. 1-1).

During 1973, SSTV provided a means by which the *U.S.S. Ticonderoga* could view the blastoff of a lunar probe. Communication between the ship in the South Pacific area and SSTV operators in the United States was conducted on 20 meters. During that time, there were less than 4000 SSTV operators around the world. Although that number has now increased, there are still many SSTV "firsts" yet to be accomplished (Fig. 1-2).

SSTV can prove indispensable during times of emergency. Pictures of damage-stricken areas may be communicated via SSTV long before commercial photographs are available.

Fig. 1-1. SSTV picture of African sunset at Z53B as received in the United States by K4TWS.

Properly used, SSTV would allow major rescue squads to be informed *before* entering a disaster area. The additional gear required to equip an amateur setup for SSTV operation need only be a camera and monitor. SSTV is used in conjunction

Fig. 1-2. SSTV picture of ZL1AOY as received by I1LC7 in Italy. This picture set a long-distance record in 1958.

with existing HF or VHF equipment. An ordinary audio tape recorder is helpful for recording pictures and playing them back at a later date.

Be prepared for the best time of your life when you enter the world of SSTV. Don't be insulted should someone react to your SSTV transmission with a remark like "is that your face or is this Halloween." Remember that before the addition of SSTV your only view was the speaker grill or receiver dial. The soft glow of a dial may have seemed nice to watch as you visualized the person you contacted, but that's 1930 style. The seventies is the era of sight and SSTV is the window of view. QSO requests turn into more of an actual visit as we look at each other's rig and family. Pet projects and schematics are exchanged over the air as we move toward more meaningful communication. The void of only audio contacts is left behind as modern day pioneers utilize SSTV to its fullest extent. One night, ZL2AAV is transmitting pictures of a live volcano, and VK5BS is showing pictures of a Tasmanian devil. The next day, CP1BCC is conducting a tour of La Paz, Bolivia (Fig. 1-3) while some others are exchanging schematics or PC board layouts. The advent of PC boards has opened a new world for the home brewer. The process of mounting parts in the proper

Fig. 1-3. SSTV picture from CP1BCC in Bolivia as received by K4TWJ in Alabama.

Fig. 1-4. This SSTV picture has been converted to fast scan via digital electronics.

holes gives everyone the capability of building sophisticated units. Circuits previously requiring months to build and a heavy technical background now become relatively short-term projects. Only a minimum technical background is necessary. The result of PC boarding is twofold: the constructor saves money when building gear, and he can acquire units not available commercially. Many SSTV operators feel that half the fun is building various circuits. SSTV is fun, exciting, and it can be relatively inexpensive.

Simply stated, SSTV is a narrow-band TV system utilizing slow-scanning rates and slightly long readout times to provide good definition in pictures. The full bandwidth of an SSTV signal is only 1100 Hz, rather than the 6 MHz of video necessary for conventional TV. (See Fig. 1-4.) This narrow bandwidth is accomplished by sacrificing an extremely small amount of definition and all movement. Movement is not necessary for amateur radio communication because we usually prefer to study every picture someone transmits—whether it is a beautiful woman or a newly built transmitter. Any loss of definition is offset by at least two major gains. We are now able to see things which would otherwise be impossible, and the monitor is usually placed near the receiver, thus a small screen is ideal and the low number of scanning lines (120 lines)

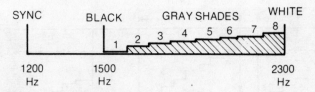

Fig. 1-5. AF analysis of SSTV signals.

is never noticed. If you doubt the validity of this statement, visit a local SSTV operator and judge for yourself. An SSTV signal is constructed with its 1100 Hz bandwidth between the range of 1200−2300 Hz (Fig. 1-5). This allows maximum stability while still consisting of audio tones falling within the range of even less expensive audio gear. Sync frequency is 1200 Hz: a "long burst" of 30 ms (milliseconds) being used for vertical reset to start each successive picture (frame) and a "short" burst of 5 ms for starting each horizontal line (Fig. 1-6). Nothing is transmitted between 1200 Hz and 1500 Hz so that video will not interfere with sync, and vice versa. Figuratively speaking, this sync is *blacker than black* so that retrace doesn't appear in the picture. Video is transmitted between 1500 Hz and 2300 Hz—1500 Hz corresponding to black and 2300 Hz corresponding to white. (See Fig. 1-7.) At least eight shades of gray are visible on most monitors, which yields very lifelike pictures (Fig. 1-8). When heard on the air, SSTV sounds similar to radioteletype (RTTY) signals except that they are more "musical" and there's a noticeable "bleep"

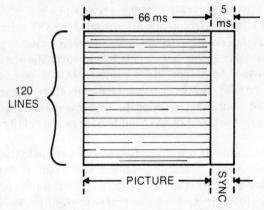

Fig. 1-6. SSTV picture time period. Next frame starts after the 30 ms long burst for vertical reset.

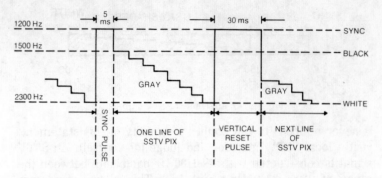

Fig. 1-7. Time/frequency analysis of SSTV pictures.

every 8 seconds. The following chart lists all present SSTV standards.

Number of lines	120
Vertical scanning rate	$\frac{1}{8}$ Hz (1 picture each 8 seconds)
Horizontal scanning rate	15 Hz
Vertical pulse width	30 ms
Horizontal pulse width	5 ms
White frequency	2300 Hz
Black frequency	1500 Hz
Sync frequency	1200 Hz

Mode of modulation for the 1200 Hz subcarrier is FM. (SSTV signals are of constant amplitude which vary in frequency according to brightness. Signals are periodically shifted to 1200 Hz for transmitting sync pulses.)

Table 1-1 is the frequency allocations for SSTV, along with suggested operating frequencies where most activity presently congregates. Notice SSTV transmissions are limited to SSB in the Advanced class portions of 80-, 40-, 20-, and 15-meter bands; however, 10-meter operation (again SSB transmitted; not AM) is open to General class and above. (There's a fair amount of SSTV on 10 meters, especially during weekends.)

As SSTV signals are audio tones, we can handle them with ease in many ways. Parallel a monitor with your receiver, and you can see pictures. Connect an SSTV camera to your transmitter's audio input and you can transmit pictures. Feed them to an audio tape recorder—then splice them together to make a running QSL show. You can even hold the tape recorder's mike

Table 1-1. Frequencies Used for SSTV in Kilohertz—A3J Emission

Legal Allocations	Suggested*
3.775 – 3.890	3.845
7.150 – 7.225	7.220
14.200 – 14.275	14.230
21.250 – 21.350	21.340
28.500 – 29.700	28.680
50.100 – 54.000	50.200

*If frequencies are busy, shift slightly up frequency to avoid QRMing SSTV pictures.

to the receiver's speaker to record pictures, or hold the transmitter's mike near the tape recorder's speaker to transmit pictures! As 2300 Hz is the highest SSTV frequency used, the frequency response of the tape recorder is not critical; however, speed regulation is a primary concern. Irregular speed will cause line-to-line jitter or "jiggled" letters (Fig. 1-9). This means the tape recorder should utilize capstan drive with a minimum of wow and flutter. Also the higher speeds, like 7½ IPS (inches per second) or 15 IPS, move tape at a more constant rate to reduce any speed irregularities.

SSTV is also creating a revolution in the home-brewing aspect of amateur radio. Old-time constructors find these AF-orientated circuits a refreshing change from RF-associated units. Newcomers and (previously) nonbuilders can use perforated boards and PC boards for constructing their first slow-scan units. (See Fig. 1-10.) This method greatly reduces any chance of error. The fact that a newcomer tries building a unit indicates his sincerity in SSTV. Should problems arise, many old-timers are always willing to assist newcomers with equipment ailments. Indeed, this status bears a very close resemblance to the early days of wireless. Amateurs across the country help each other with problems due to the pioneering efforts in this new mode of communication. Flying-spot scanners (a device for producing SSTV pictures from photos or slides, also called light-spot scanners) are a nice example of this situation. These inexpensive "first cameras" are not available commercially, thus home brewing is the only alternative. The trade-off of time for money, however, is often a strategic consideration for the newcomer.

A

B

Fig. 1-8. Some typical SSTV transmissions. (A) SSTV "roll-in" from Bolivia. Notice the dramatic effect produced from the white letters on a black background. (B) An SSTV "long-path" CQ from ZS6UR in South Africa. Signal strength was S1 with barely intelligible voice. Notice the narrow

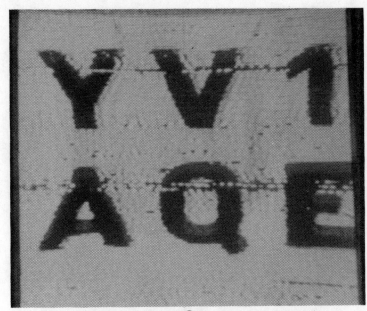

C

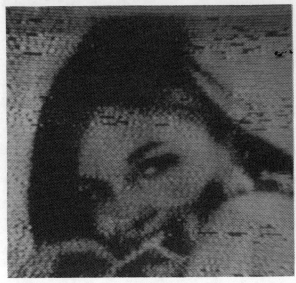

D

picture width caused by the 50 Hz African power source. (C) This transmission from South America suffered two brief bursts of noise during the 8 seconds for picture copy. (D) A weakly received SSTV picture suffers heavy interference from line noise.

A

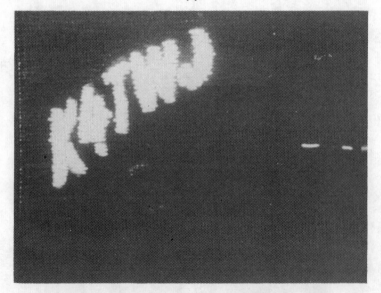

B

Fig. 1-9. Two examples of line-to-line jiggle. (A) Caused by slow tape recorder speed. This also created skewed lines that run from top left to bottom right. (B) This poor picture was caused by poor speed regulation in the tape recorder.

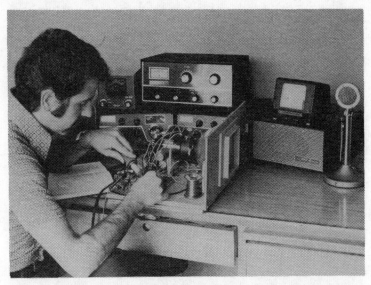

Fig. 1-10. Joe Trombino, CP1BCC, completing construction of his W6MXV monitor.

Another reason for the great amount of interest in home brewing (Fig. 1-11) is the low market for used SSTV equipment. There are many radio amateurs getting into SSTV while very few amateurs drop out of SSTV. Should one lose interest in SSTV, there are usually several others interested in purchasing his gear. The only way to retain used SSTV gear is by placing an exorbitant value on it. Soon, even the proverbial "appliance operator" begins to rightfully feel that if nontechnical friends can build SSTV gear, so can he. Soon his briefcase takes on new fillings and lunchtimes become construction periods. Finally, the SSTV gear emerges (Fig. 1-12).

One fact which can't be overemphasized is that SSTV is a simple straightforward approach to TV communication. Sophisticated video units are replaced with audio units. A knowledge of audio circuitry is the only requirement, and most amateur radio operators learned that before receiving an FCC license. If you understand how to connect a component stereo system, you can easily wire an SSTV setup. Synchronization circuits are easily designed and built because of the slow-scanning rates. In fact, a complete SSTV monitor could be built with several audio amplifiers, some extra capacitors, resistors, and a CRT (cathode ray tube); however, this

A

B

Fig. 1-11. Home-brew SSTV monitor. (A) This is what can be accomplished with a little patience. The cabinet for this unit is from a R. L. Drake TR-4. (B) Interior of the same monitor. CRT is a surplus 5FP7. Circuit is a modified W6MXV unit. (Photos compliments of WA9MFF.)

approach is considered somewhat wasteful. Anyone that considers slow scan an involved subject need only study the W7ABW oscilloscope adapter for SSTV, which appeared in the *1974 ARRL Handbook*. This adapter, shown in Fig. 1-13, is inexpensive and no more difficult to build than a transistor

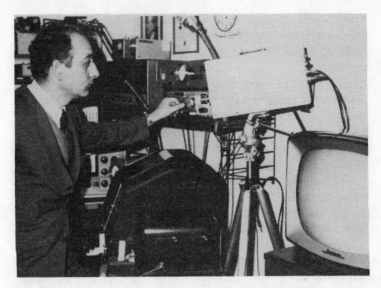

Fig. 1-12. Prof. Franco Fanti, I1LCF, adjusting the clamping level for his home-brew SSTV camera. Franco writes the SSTV column for **CQ Elettronica** and coordinates the yearly SSTV contest. He is one of the early SSTV pioneers.

radio. The oscilloscope may be replaced later with home-brew sweep circuits and CRT.

SSTV AROUND THE WORLD

SSTV interest is in no way confined to the United States. Countries in all areas of the world are realizing the tremendous fascination and advantages of visual com-

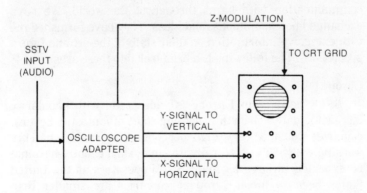

Fig. 1-13. W7ABW oscilloscope adapter pictorial diagram.

munication. Naturally, every part of the world reflects influence from regional interest and knowledge, population, natural resources, political relations, and several other variables.

SSTV is the very latest state-of-the-art evolution, which means that many countries are presently having problems accepting SSTV operations on amateur bands. Semi-isolated areas of the world often consider ham operators as spies, advocating violent overthrow of the world. These people consider SSTV as necessary as a turtle with air brakes.

There are, unfortunately, remote areas of the world suffering drought and starvation. These countries are among the DX categories. It's quite possible some of these countries have not heard of amateur radio, much less SSTV. Knowledge is carried into these countries primarily via missionaries and volunteer services; few of these individuals are hams. SSTV can fill a vital need for these areas. Natives can be taught modern farming techniques, irrigation principles, medical technology, and much more. A communications link of such magnitude can serve as a means of world assistance, to areas that are willing to learn and grow but need direction and education. One example of this situation was demonstrated when the *U.S.S. Hope* investigated the possibilities of SSTV. The ship's radio operator utilized SSTV and found it quite advantageous in communicating information to and from various areas of the world. Possibly SSTV will open many windows on the world during the next few years and help advance all mankind.

During the last few years, I have been in close communication with hams throughout the world. We have exchanged notes, opinions, and ideas. They have furnished me with extensive information on their state-of-the-art and future predictions. The following is a brief look into these areas.

Europe

SSTV activity in Europe is widespread with popularity increasing daily. Growth is nothing short of fantastic. Several countries boast a very high percentage of radio amateurs working with SSTV. Naturally, the individual countries cannot be expected to have as many SSTV operators as the United States because most European countries are smaller than many of our states.

Most SSTV circuits throughout Europe bear European design because their transistors, IC chips, and CRTs are much different from those in the United States. Don't short sell European design—it's very good. Most of these amateurs must work longer than U.S. amateurs to afford the same number of electronic components. Thus, European circuitry is 100% functional—uncomplicated but surprisingly good.

England reports issuing a large number of licenses to active radio amateurs applying to the ministry of home affairs and telecommunications for permission to transmit SSTV. Also, many other amateurs merely watch SSTV pictures. Several English amateurs have worked nearly 100 countries on SSTV.

Czechoslovakia boasts a surprisingly large number of amateurs with SSTV capabilities. Many of these enthusiasts are like American SSTV operators—they prefer to design and build gear rather than operate on the air. Also, parts availability in Czechoslovakia is a completely different situation from that in the United States. Czechoslovakian amateurs have extensive transistor resources while their IC chips are somewhat scarce. There are two leading monthly ham magazines in Czechoslovakia: *The Radio Amateur* and *Amateurske Radio*. Both publications boast an elaborate SSTV column. OK100 writes on SSTV for *Amateurske Radio*, while OK1PB writes for *The Radio Amateur*.

German amateurs are also quite interested in SSTV. Many of these DL's have designed outstanding circuitry which is beginning to be noticed as far away as Asia and the United States. Watch for a burst of video activity from this country!

Swedish amateurs design a large percentage of their SSTV gear, then work diligently to get maximum use from the equipment. The Swedish public appears quite receptive to amateur radio and SSTV. Many articles on ham communication often appear in their national public interest magazines.

Italian amateurs are well versed in the world of SSTV. This is due in part to a monthly column in *CQ Elettronica* by Professor Franco Fanti, I1LCF. Franco was one of the early SSTV pioneers, and he has contributed a substantial amount of knowledge to the subject. I1LCF and ZL1AOY set the first long-distance SSTV record. Italian SSTV operators are now working on their own versions of keyboards and titlers.

Italians are also investigating scan-conversion techniques. The limiting factor, at this time, is shift-register memory cost. One of Italy's primary public interest is fishing. Many of these fishing fleets utilize CB gear for communication needs. Slow scan is beginning to creep into these communications areas, and it's often possible to copy SSTV on 27 MHz from the Italian fishing fleets. This operation is under fire by other CB services like ambulance and taxi.

Oceania

Although a vast amount of this area is water, SSTV interest is surprisingly high. Import duties restrict the procurement of commercial gear for most amateurs. Thus most SSTV gear in Oceania is necessarily home built. Most South Pacific SSTV activity is centered primarily around the Eastern and Mountain District Radio Club of Mitcham, Victoria (P.O. Box 87). This club presently produces PC boards and kits for an SSTV monitor (termed the X51 and selling in Australia for $11) and camera (termed the X52, also sells for $11). This gear is very similar in design to Robot gear (Chapter 7) except the solid-state devices are those easily obtained in that country. The club also supplies information and exposure to amateurs interested in SSTV. One of Australia's major SSTV problems was the lack of P7 phosphor CRTs. The SSTV club realized this problem and approached a tube manufacturer for an alternative. The result was an E26 SSTV phosphor which could be inserted into any CRT envelope. It is reported that E26 phosphor SSTV pictures can be viewed in relatively bright lighting: something which was not possible with P7 tubes. E26 phosphor SSTV pictures are reddish-orange in color and lack the bright initial trace common in P7 tubes. Australia's approximate off-the-shelf cost of an 11-inch E26 tube is $30, or your existing CRT can be regunned and rephosphored for a reasonable price. Proposals have been made to change from the present 1:1 SSTV aspect ratio to the standard 4:3 ratio. This move will result in more picture area for amateurs using conventional picture tubes (and should be quite favorable to U.S. amateurs with direct-scan converters).

Digital-scan converters are almost an impossibility for South Pacific amateurs due to the cost of IC memories. This area of the world is also anxiously awaiting price drops of memory chips.

The Australian SSTV net meets Sundays around 0100Z (GMT) on 14.230 MHz. Net leaders and organizers are John Wilson, VK3LM, Barry Sharpe, VK5BS, and Doug McArthur, VK8KK. As of June, 1975 there are more than 70 amateurs in Australia with SSTV capabilities.

New Zealand SSTV activity is increasing slowly. Presently, there are approximately 20 amateurs in this small country that are equipped to operate SSTV. Poor parts availability in the South Pacific area heavily restricts SSTV progress.

There are reports of SSTV activity from Guam, the Philippines, Korea, and several other countries but this activity is basically sporadic. On the other hand, several South Pacific islands already boast SSTV activity. Two examples of this are VK9XX on Christmas Island and F08TC on Tahiti.

Asia

The Asian continent is widespread in area, ranging from Japan in the east to Jordan on the west. This geographical diversity, coupled with various languages in different areas, tends to inhibit combined advancements; however, this has not limited SSTV interest. Knowledge spreads rapidly via amateur radio bands and soon all areas of the world are involved in SSTV action.

Japan is realizing a fantastic growth in SSTV interest. This is due in part to their recent authorization to transmit SSTV on low-frequency amateur bands. There are also reports of a Japanese company manufacturing SSTV gear for amateur use in that country. Records indicate that over 100 Japanese amateurs are now authorized to operate SSTV. Most JA's exchange SSTV pictures with amateurs in Oceania and within the Far East. Possibly this situation is due to presently existing propagation.

Singapore and Hong Kong have only three or four amateurs with SSTV capabilities; however, quite a bit more activity is foreseen from this area.

Israel boasts several very active SSTV operators. These amateurs frequent the low end of the 20-meter phone band and gladly switch to SSTV when requested. Many 4X4 stations have SSTV capability—a quick inquiry during a QSO can prove fruitful. Remember that Israeli contacts count as Asia for SSTV awards.

South America

Geographic location places this continent under direct influence from North America. This is because band propagation on north—south paths is more predominant than east—west paths. Thus, South American amateurs are constantly exposed to on-the-air SSTV operations. Often, schematics are transmitted directly to South American amateurs who immediately start construction of SSTV gear.

Several South American countries constantly face revolution problems and are quite skeptic of any amateur operations. Securing permission to transmit SSTV in these countries is very difficult. Local amateurs must continually work with officials for understanding and acceptance of SSTV operations.

Columbia boasts several amateur operators that are quite active in SSTV. Many of these hams are DXers, so 20 meters is their most popular band for video work.

Venezuela is another country with high enthusiasm for SSTV. Several YV SSTV operators are also equipped to exchange SSTV via the OSCAR orbital satellite.

Bolivia revoked SSTV privileges during 1974, but several La Paz amateurs began working diligently with governmental officials to resecure authorization. This authorization should be secured presently. Rocky terrain in this area often hampers long-distance communications.

Africa

SSTV operation from this continent is somewhat sporadic, with a minimum number of native amateurs active in this new mode of communication. Expeditions and *loaner gear* have helped expose many countries to SSTV, so future growth looks very promising.

Southwest Africa has been quite active on SSTV, thanks to ZS3B. This SSTV exposure has created much interest in the southern areas. ZS6UR has also been quite active from South Africa. There are reports of growing SSTV activity from such areas as Gabon, Iran, Ethiopia, Liberia, Seychelles, and many more. The reliability of these operations should increase tremendously in the near future.

Many remote areas of Africa could benefit tremendously from SSTV use as an educational tool. Possibly future SSTV advancements will make history by performing this service.

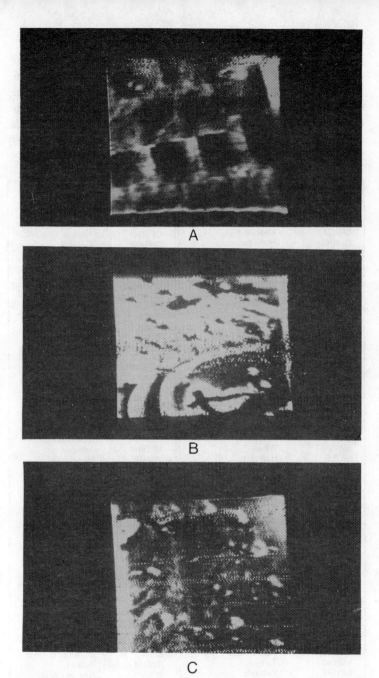

Fig. 1-14. Slow-scan shots of Mars. (A) Approaching the surface. (B) Area adjacent to spacecraft. (C) Closeup of surface showing rock erosion.

NASA Viking Visitor to Mars

There seems to be no limit to SSTV. Figure 1-14 shows the surface of Mars as seen through the camera eyes of Viking 6. Slow-scan techniques are used in both the landing craft and the orbiter. Two cameras in the landing craft sit opposite one another and pan almost 360°. Scanning time for monochrome is 10 minutes and color is 30 minutes. TV cameras in the orbiter

Table 1-2. DX List of W8YEK

C31HO	-	Andorra	LU7AAG	-	Argentina
KC4USX	-	Antarctica	LX1SK	-	Luxembourg
CN8HD	-	Morocco	OA4F	-	Peru
CP1FW	-	Bolivia	OD5HC	-	Lebanon
CR6CA	-	Angola	OE6GC	-	Austria
CL1PG	-	Portugal	OH5RM	-	Finland
CY2GB	-	Uruguay	OK1NH	-	Czechoslovakia
DJ0CN	-	Germany (1971)	ON4DN	-	Belgium
DJ7UP	-	W. Germany (1973)	OX3LP	-	Greenland
DU1FR	-	Phillippines	OY1M	-	Faeroe Is.
EA4DT	-	Spain	OZ4IP	-	Denmark
EA6BQ	-	Balearic Is.	PA0LAM	-	Neth. Amsterdam
EA8CI	-	Canary Is.	PJ2CU	-	Neth. Curacao
EL2CB	-	Iberia	PY2EEG	-	Brazil
EP2FB	-	Iran	PZ1DA	-	Surinam
ET3DS	-	Ethiopia	SM4AMM	-	Sweden
F6AXT	-	France	SV1CG	-	Greece
FG7XT	-	Guadeloupe Is.	K4PGM/TI7	-	Costa Rica
FL8BH	-	Fr. Somaliland	TJ1AX	-	Cameroon
FM7WW	-	Martinique	TR8WR	-	Galion Rep.
FP0AO	-	St. Pierre Is.	TU2DO	-	Ivory Coast
G5ZT	-	England	VE6RM	-	Canada
GC3YIZ	-	Guernsey Is.	VK5MF	-	Australia
GI3WWY	-	North-Ireland	VK9XX	-	Christmas Is.
GM3KJF	-	Scotland	P29MC	-	New Guinea
GW3DZJ	-	Wales	VP2AR	-	Antigua
HA7LF	-	Hungary	VP2ME	-	Montserrat
HB9IT	-	Switzerland	K4GXO/VP7	-	Bahama Is.
HB0NL	-	Liechtenstein	VP9GR	-	Bermuda Is.
HC1BU	-	Ecuador	VO9R	-	Seychelles Is.
HK7XI	-	Colombia	VS6AI	-	Hong Kong
HL9WI	-	Korea	VU25KV	-	India
HP1XMU	-	Panama	XE1JM	-	Mexico
HR2HH	-	Honduras	XW8AX	-	Laos
HS1AHE	-	Thailand	YN3RBD	-	Nicarauga
HZ1SH	-	Saudi Arabia	YU2DCS	-	Yugoslavia
I1LCF	-	Italy	YV5AS	-	Venezuela
IS1PEM	-	Sardinia	ZF1AO	-	Cayman
JA7FS	-	Japan	ZL1AOY	-	New Zealand
JY8AA	-	Jordan	ZS6UR	-	South Africa
W4MS	-	United States	ZS3B	-	South West Africa
KC4DX	-	Navassa	4X4VB	-	Israel
WA6AXE/KG6	-	Guam	5W1AT	-	West Samoa
KH6DEH	-	Hawaii	6Y5PB	-	Jamaica
KL7DRZ	-	Alaska	YB3AAY	-	Indonesia
KP4GN	-	Puerto Rico	8RIW	-	Guyana
KS6DW	-	Samoa	9K2AM	-	Kuwait
KV4CM	-	Virgin Is.	9Q5BG	-	Rep. of Congo
KX6DR	-	Marshall Is.	9X5PB	-	Rwanda
LA3SG	-	Norway	9Y4VU	-	Trinidad

scan at about 4.5 seconds. Photo A in Fig. 1-14 was transmitted as the craft approached the landing site. The white square near the center of the photo was the landing site. This picture was received on 14.230 MHz during July, 1976.

Photos B and C were taken after the landing. The black object at the bottom right in photo B is the shadow of the spacecraft's leg. Top portion of the shot reveals windswept soil on the surface. The circular patterns are due to impact. Rock formations are illustrated in photo C. There is some QRM in this photo, but enough detail remains to see rock erosion.

SUMMARY

The previous analysis primarily considered worldwide SSTV interest for popular countries with outstanding SSTV operations. It must be realized that each of the world's 300 countries cannot be discussed individually.

Don't jump to the conclusion that SSTV interest is small. There are many small countries where SSTV interest is flourishing. DXers can appreciate the supreme challenge of SSTV—work 100 countries on video, then try for 5 bands DXCC/SSTV.

The best evidence of worldwide SSTV interest is W8YEK's recent accomplishment of DXCC on SSTV. After many months of devoted SSTV operations, Gene successfully completed two-way video exchanges with hams in over 100 countries! Table 1-2 lists the stations and countries he contacted and can serve as a guideline for DX-minded SSTV operators.

Chapter 2
Understanding SSTV Gear

Like any type of electronic equipment, SSTV gear may be considered from either the standpoint of technical design or the overall operational standpoint. As this chapter is written for the semitechnically oriented amateur or SSTV newcomer, it will consider SSTV equipment from the operational standpoint, using block diagrams. This will allow sufficient understanding of all SSTV gear regardless of its active or passive filters, or slope discriminators.

SSTV circuitry (Fig. 2-1) is relatively noncritical and easy to understand because of the associated low frequencies. Audio amplifier "perforated boarding" is usually sufficient for wiring. Audio chokes and simple toroids replace critical bandpass filters and HF video peakers. (See Fig. 2-2.) The *basic* operation of all SSTV monitors is similar, therefore a simplified block diagram (Fig. 2-3) and description follows.

SSTV MONITORS

SSTV signals from the receiver's earphone jack or speaker terminals first enter a conventional limiter/amplifier where amplitude variations (QRM, QRN, etc.) are reduced. This amplifier merely boosts the SSTV level so it is easier to use. Limiting action may be accomplished by simple diodes or transistor circuits. We now have a high-level constant-amplitude signal for application to the frequen-

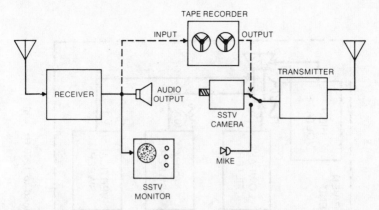

Fig. 2-1. Basic SSTV interconnections with the HF station. Tape recorder can be added to permit relay-station operation or to keep received transmissions for future reference.

cy-sensitive circuits to retrieving sync and video information (point 1 in Fig. 2-3). The sync discriminator is sharply tuned to 1200 Hz so only SSTV sync pulses will pass on to the sync amplifier (point 2 on the block diagram). This discriminator extracts sync only by using filters. While some monitors use Butterworth filters, some toroids, and others tunable active filters, the overall objective is the same—to pass 1200 Hz sync

Fig. 2-2. An SSTV monitor goes together at WN6EKO. Power supply is on the left, circuitry is in the middle, and CRT is on the right. When interconnected, this unit provides an excellent SSTV monitor. Notice the simplicity of construction—merely mount the parts on the PC board.

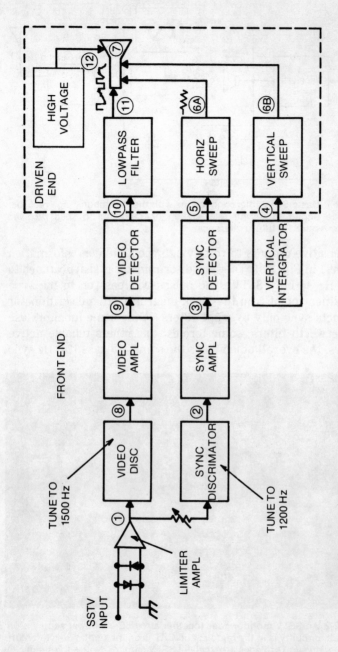

Fig. 2-3. Block diagram of a basic SSTV monitor. The sync discriminator is tuned to 1200 Hz, while the video discriminator is tuned to 1500 Hz.

pulses only. A full discussion of bandpass filters is presented in the *SSTV Handbook* by *73 Magazine*, so that information will not be repeated here. After amplification (point 3) and detection, the vertical pulses (point 4) are separated from the horizontal (point 5) pulses. Both pulses may now drive their respective sweep circuits (point 6A and 6B). The sweep circuits cause a raster to be painted on the CRT screen by the electron gun (point 7). During the same time period, the incoming SSTV signal was also being fed to the video discriminator (point 1). This circuit is tuned to pass only 1500—2300 Hz. This band of frequencies is then slope detected, producing low output for 2300 Hz and high output for 1500 Hz (point 8). Some variations in video detection methods will be noted in monitors, however, the basic idea is the same. This voltage is then amplified (point 9), detected (point 10), run through a low-pass filter (point 11), and applied to the CRT for intensity modulation (point 12). As an overall result, we have light and dark variations placed at appropriate points on the screen to reproduce the SSTV picture. Basically, this describes the theory of operation behind all SSTV monitors. Under S9 (solid copy) conditions (called closed circuit among SSTV operators) all units will produce pictures of equal quality; however, QRM will cause interference to SSTV pictures. This is where the more sophisticated units excel. Naturally, the more elaborate a monitor (more bandpass filters, timing, gating, level comparison circuits, etc.) the more involved and expensive the unit. Just remember, stripped of its frills even exotic monitors are simple.

HOME BREWING SSTV MONITORS

Construction of a home-brew SSTV monitor will take careful preparation, experimentation, and layout on your part before you can successfully announce to the world that you are *on the air*. Figure 2-4 shows a carefully planned vacuum-tube type SSTV monitor by I1LCF. The mini-3FP7 CRT is mounted in the center of the chassis, which helps separate the sweep circuitry from the signal circuitry. Notice the care that was taken to align the components vertically and horizontally. The CRT is shock mounted to prevent breakage, the anode lead is mounted well away from the PC board to avoid arcing. Figure 2-5 shows a home-brew solid-state version of an SSTV monitor with six PC boards to provide a modular approach to SSTV.

Fig. 2-4. The I1LCF SSTV monitor with the 3FP7 CRT mounted in the center of the chassis.

Employing semiconductor circuits allows the constructor to miniaturize his project and provide that commercial look. Using IC (integrated circuit) devices carry the miniaturizing process even a step further. Of course, neither one of these examples is a first-attempt project.

FLYING-SPOT SCANNERS

Although flying-spot scanners are not available commercially, they make an ideal first-camera project for the beginning SSTV enthusiast (Fig. 2-6). Flying-spot scanners are easy to build and yield very good quality pictures. Circuitry is straightforward because a CRT is used to scan still pictures directly at slow-scanning rates. The long, light-tight box employed to house the flying-spot scanner may be used in a dark room. This is often an advantage over the SSTV camera which requires bright lights on subject matter for proper results. A brief explanation of flying-spot scanner operation follows. (See Fig. 2-7.)

Horizontal and vertical sweep is applied to a CRT, thus producing a raster (point 1 on block diagram). No video modulation is applied to this CRT, so the resulting raster is

A 1962087

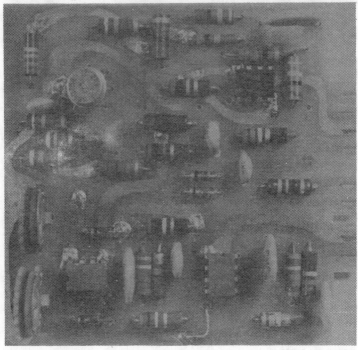

B

Fig. 2-5. Home-brewed WA9MFF SSTV monitor. (A) Home brewing an SSTV monitor is simple once you understand the basic principles of operation. (B) One of the six PC boards used in the WA9MFF monitor. Notice the functional layout and the modular design. The key to construction is don't overcrowd PC boards.

Fig. 2-6. Business end of a flying-spot scanner. Each slide is positioned against the CRT. The PMT is located at the far end of the box. The deflection circuits are located in a separate cabinet.

completely white. The light from this raster is blocked out at various places corresponding to the picture (transparency) placed in front of the CRT (point 2). The varying light intensities now fall on a light-sensitive photomultiplier tube (PMT), producing a very small video output voltage (point 3). This video voltage is amplified (point 4) and applied to an SSTV master oscillator, producing a resulting 1500–2300 Hz video swing (point 5). This SSTV oscillator is also periodically shifted down to sync frequency (1200 Hz) to produce the composite SSTV output. The sync generator which sends sync shift pulses to the master SSTV oscillator (point 6) also synchronizes the start of each scan on the CRT (point 7). The overall result is a video frequency between 1500 Hz and 2300 Hz being placed in the proper picture location with relation to each line's beginning sync pulse of 1200 Hz—a high quality SSTV picture results.

SSTV CAMERAS

The first SSTV cameras (Fig. 2-8) utilized specially produced slow-scan vidicons or TV station Plumbicons, which were pulled from service due to minute imperfections. Ordinary vidicons produced too much noise in the pictures

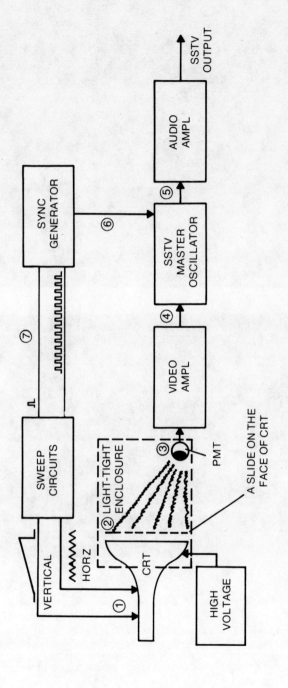

Fig. 2-7. Block diagram of a flying-spot scanner.

Fig. 2-8. Home-brew SSTV camera (I1LCF). (A) This half of the I1LCF unit contains the camera with the video amplifier circuitry in the lower portion of this photo. (B) Support circuitry for the picture-pickup portion of the camera. The power supply is housed in another cabinet to eliminate hum.

when operated at the slow-scanning rates. This is because the charge stored on the target diminishes between scanning-beam visits to each picture element. If video information is not present, then only noise is left. Obviously a more acceptable SSTV generation method was required, so sampling cameras gained popularity. Most present SSTV cameras use sampling techniques with the latest trend being direct scan conversion.

Basically, sampling means that a fast-scanning rate is employed for the camera pickup tube, and portions of this fast-scan picture are used to make up the SSTV driving signal. This driving signal then varies the frequency of an SSTV master oscillator (1500−2300 Hz). Naturally, this oscillator is periodically shifted down to sync frequency (1200 Hz) by the camera's sync generator. Although additional support circuitry is necessary, this describes the general concept. The above procedure permits a TV camera tube (like a vidicon) designed for a fast-scan rate to be fully utilized in a slow-scanning system. The sampling is accomplished by taking only a portion of each fast-scan frame. This portion is stored, and the time between samples is used for slow-scan transmission of the sample. In the case of SSTV, it is convenient to take one line of slow-scan video from each fast-scan frame.

In a sampling camera, the camera tube (vidicon) operates at a fast-scan rate. It is often convenient to provide a closed-circuit monitor operating at this scanning rate. This monitor is refreshed at the fast-scan rate, thus it presents a bright motion picture. Such a fast-scan monitor is a convenient asset when initially setting up the slow-scan camera.

As previously mentioned, it is convenient to take one slow-scan line from each fast-scan frame. This format is illustrated in Fig. 2-9. Note the fast-scan lines are oriented vertically while the slow-scan lines are horizontal. The number of scans per second is given for each direction of the camera and monitor. The camera and monitor scans are kept in step by the sync pulses. Both units start their scans at the upper left corner. When the scans pass the point marked 1, a short sample of the camera signal is taken and stored for transmission. The stored sample is displayed at 1 on the slow-scan monitor. When point 2 is reached on the camera tube, a new sample is taken and the stored value is changed to

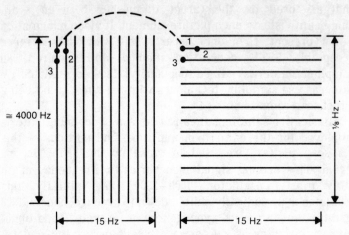

Fig. 2-9. Fast- to slow-scan sampling format.

represent the scene intensity at that particular point. While the camera scan was traversing an entire line from 1 to 2, the monitor scan, traveling slowly, moved to the adjacent picture element. Thus each line of the camera frame is sampled once to create one line of the slow-scan display. With the beginning of a new frame in the camera tube, the sampled spot has moved down to 3 and a new line is started on the monitor. This process continues until the entire monitor is covered with picture information. The sampling function in the camera is accomplished by a circuit which times the occurrences of the short samples and another circuit which captures and stores and samples.

SAMPLING CAMERAS

Now let's consider a basic sampling camera in block diagram form, similar to the way we studied an SSTV monitor. A sampling camera diagram is shown in Fig. 2-10.

First, the picture information is picked up by the vidicon. This video is output at a fast-scan rate (point 1). This video signal is then amplified and limited so black and white excursions will be within specified parameters (point 2). The previous procedure gave a high-level video signal that can be utilized without destroying its content. Some cameras have adjustable limiters so black and white extremes may be varied. The processed fast-scan video is now applied to a hold-

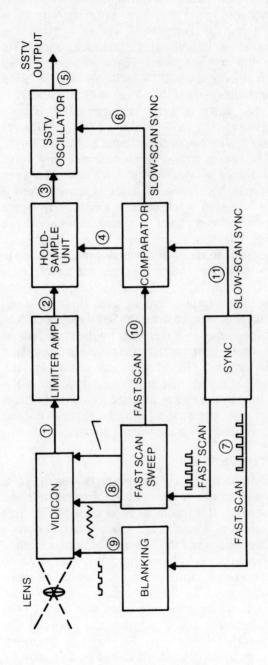

Fig. 2-10. Block diagram of a sampling camera.

sample unit which incorporates a timing circuit for standby and output functions (point 3). This sampling circuit is externally keyed by the comparator (point 4), thus the comparator is directing the hold-sample and output-sample (to drive SSTV oscillator) functions. These slow-scan video samples (point 3) then shift the SSTV oscillator between black and white frequencies (point 5). The SSTV oscillator is periodically shifted down to sync frequency (1200 Hz) by applying a fixed bias to the SSTV oscillator. This operation is usually controlled by the comparator (point 6), so SSTV sync will fall during fast-scan sync—not when samples are being taken from the fast-scan video. Fast- and slow-scan sync pulses are produced by frequency dividers driven from a master oscillator locked to 60 Hz (point 7). Fast-scan rates are used to sweep the vidicon (point 8), blank vidicon retrace (point 9), and drive the comparator (point 10). Slow-scan rates are used to drive the comparator for samples (point 11) and to sync the SSTV oscillator (point 6). The end result is a slow-scan picture produced from *bits* or samples, of the fast-scan picture. The sampling circuit holds these bits until the comparator directs them to drive the SSTV oscillator. The SSTV oscillator frequency is determined by the amplitude of each sample. This sample and output operation is happening at such a rapid rate that an SSTV driving signal results. Briefly, and in the simplest of terms, that describes the operation of a slow-scan sampling camera. The other SSTV method, direct fast-scan to slow-scan conversion, involves using a regular closed-circuit surveillance camera without any modifications.

DIRECT SCAN CONVERSION

Since its inception, the idea of converting slow scan to fast scan and vice-versa has been foremost in the minds of those pioneers interested in the technical aspects of SSTV. This system would consist of a small "black box" placed between the fast-scan video source and the SSB transmitter for sending SSTV. The video source could be an inexpensive surveillance fast-scan camera, video tape recorder, or even the video output from a commercial TV studio. Reception of SSTV could be accomplished by placing another black box between the communications receiver and a conventional fast-scan TV. Absolutely no modifications to the TV would be required. Incoming slow-scan pictures are then viewed on a regular TV in bright long-lasting form. The previously described techniques are a revolutionary approach to SSTV, and they are presently somewhat expensive compared to conventional SSTV transmission and reception methods.

One of the first methods of scan conversion involved a special tube which functioned like two vidicons mounted face to face sharing a common target. This unit is illustrated in Fig. 2-11. One end of this tube acted like a CRT projecting a slow-scan picture onto the target which stored the image in capacitive form.

The other end of this tube then scanned the stored image to output with fast-scan TV. The project/scan function could be reversed for fast- to slow-scan conversion: a fast-scan signal would drive the *projection end* while the *scanning end* fed conventional SSTV circuitry. Unfortunately these scan-conversion tubes were expensive and difficult to locate. Thus other methods of scan conversion were sought.

Today's most popular method of scan conversion utilizes digital electronics techniques. This scheme is presently rather expensive compared to conventional slow-scan transmission and reception tactics; however, the cost of MOS (metal-oxide semiconductor) shift registers may drop substantially in the very near future, making digital conversion practical for everyone. While scan-converter circuitry varies between individual units, the basic operational concepts are the same. We will now discuss scan conversion from a simple standpoint using block diagrams for clarity. An in-depth study of digital

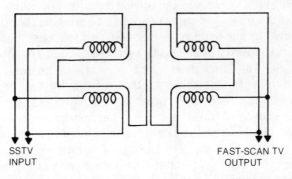

Fig. 2-11. The scan converter tube was one of the first methods used for direct scan conversion. It functions like two vidicons face to face, sharing a common target.

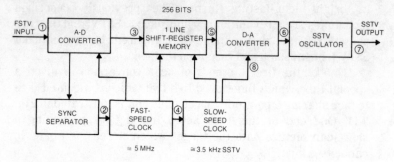

Fig. 2-12. Digital fast- to slow-scan converter block diagram. Memory load and output functions are controlled by special clocking circuits.

scan conversion techniques, including logic circuit design, will be given later in this book.

DIGITAL FAST- TO SLOW-SCAN CONVERSION

The fast- to slow-scan converter is basically a line converter; it sequentially loads each line of fast-scan video into the shift-register memory, then extracts this information very slowly. Since the shift register will not accept various levels of voltage, suitable converters must be placed before and after the shift register. Memory load and output functions are controlled by special clocking circuits. A block diagram of this system is shown in Fig. 2-12.

When fast-scan TV is first applied to this unit, one line of video information loads the analog-to-digital converter. This is shown at point 1 on the block diagram. This A-D (analog to digital) unit employs OR gates and a series of voltage level comparators to change the analog video voltages into some form of binary-coded equivalent. A fast-scan sync pulse is then extracted by the sync separator and used to trigger operation of the fast-speed clock. This is indicated at point 2 on the block diagram. The fast-speed clock now allows digitized video information to move from the A-D converter into the shift-register memory (point 3). This operation takes approximately 1/15750 second. The fast-speed clock now clamps all information flow into the shift register, and the loading process stops. The fast-speed clock now triggers the slower clock (point 4) and the unloading process begins. The slow-speed clock allows digitized video information to slowly move from the shift register into the D-A (digital to analog) converter (point 5). This conversion is accomplished by a series of transistors and resistors connected so that various voltages are produced by each binary count moving into the converter. Output voltages are shown at point 6. Voltage levels from the D-A converter are used to frequency modulate the SSTV oscillator between black and white frequencies (point 7). At the end of the slow-scan line, a sync-level pulse goes to the D-A converter (point 8), and the line conversion process is now repeated; the next fast-scan line is quickly moved into the shift register and drawn out very slowly. Approximately 120 of the previously described operations occur during an 8-second period, after which a full SSTV picture evolves.

One must realize the previous explanation concerns scan conversion in its simplest form, stripped of all extras. An expanded description is given in Chapter 6.

DIGITAL SLOW-TO FAST-SCAN CONVERSION

We will now discuss one of the most outstanding SSTV innovations to date—the slow- to fast-scan converter (Fig. 2-13). Although this unit is more sophisticated and expensive than conventional gear, it appears to be spearheading a complete new era in SSTV monitor design. Bearing in mind that Chapter 6 is devoted solely to scan-conversion techniques,

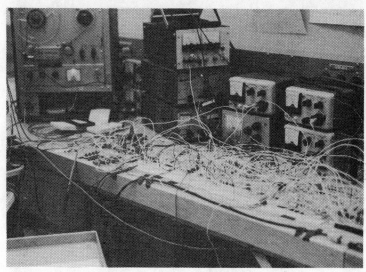

Fig. 2-13. Breadboard stage of a slow- to fast-scan converter. Ever wonder what a slow- to fast-scan converter looks like during its development? This one actually works! Many IC chips are buried under those wires. The next step is PC layout.

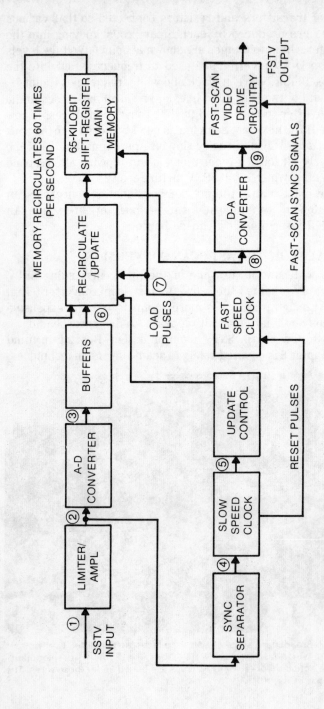

Fig. 2-14. Block diagram of a slow- to fast-scan converter. Notice that there is an A-D converter and D-A converter.

we will consider the simple block diagram analysis in Fig. 2-14.

Incoming SSTV is first applied to a limiter stage for securing a constant amplitude signal which can be applied to several circuits. This is point 1. The slow-scan video is then fed into a somewhat modified A-D converter (point 2), which produces a binary-coded equivalent for each specific level of SSTV video input. At the end of each slow-scan line, the digitalized video information sequentially moves into the appropriate buffers. This is point 3. After approximately 120 of these loading operations, a slow-scan vertical-reset pulse is received and extracted by the sync separator (point 4). This pulse directs the slow-speed clock to trigger the update control (point 5). The update control allows each buffer to sequentially unload its contents into the main memory (point 6). Operation of the 65-kilobit shift-register memory is being controlled by the fast-speed clock (point 7). Information stored in this memory will recirculate, then output at a fast-scan rate. New or incoming slow-scan information is used to replace old data in this fast-running "rat race" for converting each successive SSTV picture. Fast-scan digitalized video is now fed into the A-converter (point 8). This unit takes digital-coded information from the main memory, converts it to ordinary binary code, and produces an analog voltage output (point 9). This voltage drives fast-scan interface circuitry so that the scan converter can connect to a TV via the antenna terminals.

Chapter 3
Setting Up the SSTV Station

As previously mentioned, SSTV signals are made up entirely of audio tones, permitting a simple plug-in arrangement to your existing gear. The SSTV monitor parallels across the receiver's speaker while a SSTV camera or tape recorder feeds the transmitter through a simple switching and impedance-matching arrangement. Many SSTV newcomers start with only a tape recorder and some SSTV tapes recorded by a fellow SSTV operator—camera and lights may be added to the setup later. In fact, the SSTV camera and lights can be a group endeavor among local SSTV operators which could be shared on a *loaner* basis. Refer to Fig. 3-1 for an idea of what a well planned functional SSTV layout might look like.

If one tape recorder is used, a switching arrangement (Fig. 3-2) is necessary for switching between inputs from the receiver or camera and outputs to the transmitter and monitor. While commercial manufacturers include these circuits in their gear, the home brewer usually designs a switching system for his particular needs. Basically, a system should be designed for accomplishing all functions without unplugging and swapping cables. Further, the receiver's audio should not "recycle" to the transmitter's input, and the monitor should view both received pictures and transmitted pictures. Those preferring a more flexible arrangement might consider using two tape recorders; in addition to eliminating

Fig. 3-1. Some well planned SSTV layouts. (A) This functional SSTV layout is by W6KZL. A Robot SSTV camera is to the far left, 10-inch SSTV monitor is in the top center, and a flying-spot scanner plus a 3-inch SSTV monitor are below. A cassette tape recorder is beside the 3-inch monitor. The communication gear is behind Glen. (B) This compact but efficient SSTV setup is WB2GUB. Camera is on the left, monitor is at the top, and tape recorder is at the bottom.

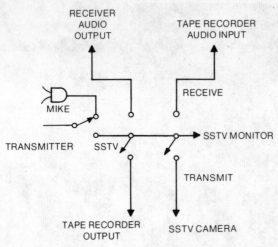

Fig. 3-2. Simplicity is the keynote in this basic video switcher of K4TWJ.

reel or cartridge swapping between transmit and receive functions, this acts like an additional video source when producing programs. For example, station identification (ID) could first be recorded on one tape recorder, then added before and after "live" shots being recorded on the other recorder. These three *recordings* may then be transmitted anytime desired, while the camera is still free for additional shots if preferred. Special consideration should be given to monitor and tape recorder location, as their wiring will probably run around or under the desk to the switching and lighting arrangement. Remember the monitor should be easily visible while making camera adjustments. Floodlights should be positioned to exclude screen illumination if at all possible. The time spent planning your SSTV layout will repay you manyfold in operating enjoyment and comfort, whereas a poorly planned layout will really make you work during those contacts! Since SSTV programs are usually made up when frequency bands are dead or activity is low, the program tapes and tape recorder location is quite important when considering on-the-air operating comfort. Taped programs are definitely preferred for CQs, IDs, and other *often used* pictures, as they provide effortless SSTV operating.

MONITORS

When first using an SSTV monitor, one is tempted to advance the contrast and brightness controls to the point of over clarity. As a result pictures will be quite contrasty, expecially on IDs, but will lack full gray scale. If only a reasonable amount of contrast, supplemented with the necessary brightness, is utilized, maximum gray-scale rendition will be achieved (Fig. 3-3), resulting in more lifelike reproduction of scenes.

Taping, experimentation, and comparison should prove the best setting of these controls for particular taste. Remember, the contrast control varies the video amplifier gain (how much "cutoff bias" voltage for CRT is developed by the incoming signal), while the brightness control determines screen intensity (electron flow from gun) without modulation from incoming video. (See Fig. 3-4.) Commercial gray-scale tapes are helpful for these adjustments as they swing from reference black (1500 Hz) to peak white (2300 Hz), whereas off-the-air signals may vary from only 100—2100 Hz. This is further influenced by tuning of the SSB signal. Another point on new monitors is picture position and focus. While proper centering often requires a slight adjustment of either the horizontal- or vertical-centering pots, a "tilted picture" may

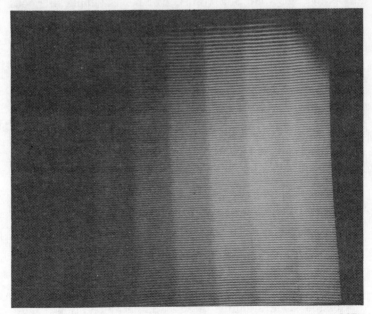

Fig. 3-3. Eight-shade gray scale. This displays properly adjusted SSTV monitor. Photo was shot from a Venus SS2 monitor. (Courtesy Venus Scientific Inc.)

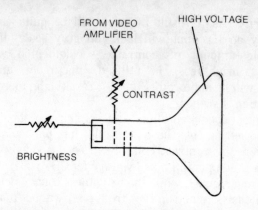

Fig. 3-4. This is how the SSTV video controls the CRT electron gun output.

also be corrected by gently rotating the yoke on the CRT neck. (Be careful to avoid the high-voltage connector on the CRT bell. If skeptical, hold one hand behind your back and make adjustments with the other hand. This way it's extremely difficult to get across the high voltage, although most SSTV units do use solid-state pulse high-voltage supplies that are less dangerous than SSB transceiver power supplies.) Focus should be sharp enough so you can see between the lines in a picture when inspecting closely (see Fig. 3-5). If the lines appear to touch each other, a slight adjustment of the focus pot or magnet assembly (depending on the particular unit) should improve definition. A black poster-paper hood around the monitor screen might prove advantageous during high room-illumination periods.

CAMERAS

SSTV cameras require slightly more consideration than monitors when setting up, due to lighting, focusing, and electrical adjustments. Ideally, camera and floodlights should be positioned to televise either the operating position or swung around toward a studio setup to view printed material for titling, IDs, etc. This gives maximum use of the gear with minimum effort, which is a definite advantage when you consider all operations involved—taping, focusing, changing material, varying illumination, and time to enjoy it. A reasonable understanding of SSTV camera operation is quite advantageous before actual operation is attempted. Although home brewers usually acquire a vast knowledge before and during construction of SSTV gear, it is suggested the newcomer read thoroughly the camera's instruction manual beforehand. Usually the contrast control determines the SSTV master oscillator swing, while the brightness control establishes the oscillator's upper and lower limits. There is a small interaction of these controls in most cases as they both are varying voltages on, usually, a multivibrator. This is easiest visualized when considering a multivibrator with three pots in series with its voltage source. One pot varies the applied voltage (frequency limits), while the others determine the variations (swing) within these limits. Further, consider one of these *variation* pots as the incoming-video signal source, and you have a general idea of the operation. Again, reading your particular unit's instruction manual is the keynote.

One of the more important controls in a camera is the beam-current control. A brief description of camera pickup tube (either vidicon, Plumbicon, or orthicon) operation is in order to fully understand this control. The charge-storage effect of an iconoscope will be exemplified for simplicity. A

Fig. 3-5. A clear SSTV picture, properly focused, shows excellent definition and sharp individual lines. High resolution is quite common under good signal conditions.

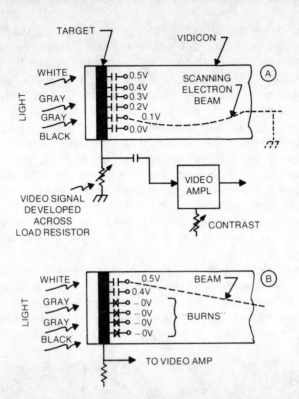

Fig. 3-6. Camera pickup tube operation. (A) The target consists of several thousand tiny light-sensitive globules that store voltage. Load resistor is effectively connected across the capacitive globules. The voltage charge of each globule varies from +0.5V (white) to 0V (black). (B) If a spot is burned on the target, the voltage output of the burned globules is 0V, showing black instead of the desired gray.

light-sensitive target is used in the front end of this tube for converting the televised scene into electrical voltages equivalent to the picture. (Refer to Fig. 3-6.) The target consists of several thousand tiny light-sensitive globules which can store this generated voltage in a capacitive manner for a short time. Each globule thus has a different voltage across it, depending on the illumination of that particular area of the scene. All of the globules have *one side* connected to the target connection and target load resistor. The ground end of the target load resistor is connected to the *other side* through the scanning electron beam, thus developing the output video signal. The globules will now produce maximum voltage output for white areas of the scene and no voltage output for dark areas of the scene. Should the scanning beam inadvertently stop while scanning the target, it will burn the target and destroy the globules it was touching. Thereafter, there will be a black spot on all televised pictures where these globules were destroyed.

Now consider a case where the beam current is excessive. A raster will be burned into the pickup tube, continually reducing its performance and destroying the weaker areas. As a result, tube life is short and scenes are improperly reproduced. A fire hose "scanning" a billboard is a good analogy here. If the water stream (scanning beam) is too forceful it will knock over the billboard. Further, water will splatter onto dry (equivalent to dark) areas, causing an excess of water (improper illumination) when scanning the full billboard. Should the hose stop scanning, it will *punch* through the billboard. If you drop the hose pressure, you can equally dampen the full surface of the billboard without damaging it. In other words, use *just enough* beam current for thorough scanning of the target, and increase target gain for maximum camera sensitivity. Unlike a TV picture tube, you cannot directly see the raster being scanned in the camera pickup tube.

It is possible for an SSTV newcomer to adjust his camera and monitor improperly, producing good pictures in his shack but indistinguishable pictures when received by others. Usually this is created by too much contrast on the monitor, then adjusting the camera while watching this monitor. Remember, the camera's audio output must swing from 1500 Hz at black to 2300 Hz at white. Try paralleling an earphone or speaker across the camera's audio output to listen for the frequency swing when you cap or uncover the lens. You might borrow a friend's audio oscillator to tape record these 1500−2300 Hz tones for your test. Does it sound like other SSTV signals you hear on the air? Keep adjusting until you get the required audio swing before disconnecting the earphone. Taping some good quality pictures off the air, then adjusting your camera to equal these *without* readjusting monitor brightness or contrast controls usually proves fruitful also. Now that you have the camera and monitor operating properly, it's time for action, so let's get started!

Try placing a full page of printed material in front of your camera for focusing adjustments. This has the advantage over pictures in that adjustments can be realized immediately, rather than 8 seconds later. Focus/distance calibrations on the lens will give a rough approximation of adjustment starting points.

A variable light dimmer is a real gem when adjusting illumination for various subject matter encountered. You may quickly find this a more suitable method than changing f/stops constantly. The use of bright lights and larger number f/stops (smaller opening) gives the best results, as subject material is fully illuminated, producing a solid picture on the camera's pickup tube target. Remember—cap your camera lens when not in use so light will not reach the sensitive target.

The choice of lenses for your SSTV camera is quite varied, yet relatively simple. For all practical in-shack SSTV operation, the popular 19—25 mm lenses are excellent. These usually focus from 10 or 15 feet down to a couple of inches. Usually, f/stops range from around f/1.4 to approximately f/22, which performs quite well with 150—250W light illumination of the subject. Although not an absolute necessity, a closeup lens is quite handy for detailed work, whereas a 200 mm or larger (telephoto) lens is suitable for long-distance shots. Generally, these long-distance shots require more light than closeup pictures, thus you get into larger sized lenses. The light gathering property of a lens is directly proportional to its f/stop. Usually each click of the f/stop doubles or halves the previous amount of light gathered, thus you should choose a lens with at least f/1.9 as a minimum, permitting use of ordinary household lights (70—200W) and reflectors. All commercial slow-scan cameras are designed to accept the popular C-mount screw-type lens, thus you might consider letting that extra lens on the photographic camera pull double duty on the SSTV camera from time to time for special productions. There are several methods of accomplishing closeup shots. One method is with the inexpensive closeup adapter lens fitted over your regular 19—25 mm camera lens; these are usually available in photographic supply houses and are a form of magnifying glass for the lens. Another method of accomplishing closeups is increasing lens-to-camera pick-up-tube distance. This may be accomplished by moving the pickup tube away from the lens on a form of sliding track. An adapter of this type is available for the presently popular Robot camera by an independent company. The unit will permit regular-size postage stamps to completely fill a raster while using an ordinary 25 mm lens on the camera. These same results may also be obtained by increasing the distance between lens and camera. The easiest method of accomplishing this is an extension tube, or barrel, that screws into the camera. Place the lens into the other end of the extension tube. This method boasts the advantage of a screw-in adapter which can easily and inexpensively be fabricated from available items around the shack. Experimenting with various lenses often proves to be a favorite pastime among many SSTV operators.

Bright lights are not a prerequisite for slow-scan operation as exemplified by the following slide projector technique. Possibly you have noticed the inexpensive battery-operated slide viewers available in most variety stores. When a slide is placed in the unit, voltage is applied to a small lamp for illumination of the slide for viewing (See Fig. 3-7). If one of these units is placed in close proximity to the camera lens (almost touching it) the slide will exactly fill a slow-scan raster. Illumination is perfect and the camera's f/stops will

Fig. 3-7. The K4TWJ slide chain is ideal for televising 35 mm slides or 127 negatives. An internal video inverter on the camera produces positive picture output. After the station ID, the camera can be swung around to view the studio area.

usually be in midrange when proper operation is acquired. Not only are slides available of cities and areas of interest, but you can make your own slides inexpensively by purchasing a bulk quantity of slide blanks and loading them with home-brew transparencies. This may be accomplished with a sheet of clear plastic and a felt-tip pen or dry-transfer lettering, available at office supply outlets. The inexpensive 126 Instamatic film is exactly the right size to fill these slide blanks. A roll of 12 exposures may be shot and developed into negatives quite inexpensively. When these negatives are placed in the slide viewer and the camera switched to *video invert*, slow-scan pictures may be accomplished quite inexpensively to fill a variety of needs. The slide projector may be placed on any convenient object, so that the camera may be swung from its usual position to "look into" the slide viewer. Those desiring a more elaborate system for viewing Polaroid pictures directly might consider fabricating an adapter that could snap on to the SSTV camera's front, then front-illuminate photos with small pilot lamps. One reminder while on the subject of SSTV cameras and photo viewing arrangements—placing material on a table, then having the camera look face down on this is not recommended. Any minute impurities in the camera pickup tube fall on the target, producing damaged (black) spots.

Another popular device for transmitting pictures, again even in a darkened room, is a flying-spot scanner (see Fig. 3-8). The flying-spot scanner should not be overlooked by the SSTV enthusiast, as it provides top quality pictures without the need for a camera. And you have the enjoyment of building it yourself. There are presently no manufacturers of flying-spot scanners, but they may prove to be a worthy addition to any SSTV setup.

An often raised question is "How can you match SSTV audio output to transmitter audio input?" Several methods are possible here, ranging from transformer circuits to attenuator networks. One of the simplest is the two-resistor network shown in Fig. 3-9. This scheme works quite well and resistor values are not critical; lowering specified values slightly will give more audio drive to the transmitter or tape recorder, while maintaining the same setting of SSTV output circuit. This scheme is handy for matching the camera to the tape recorder or transmitter and also tape recorder to transmitter.

A

B

Fig. 3-8. Flying-spot scanners require no external light to televise pictures. (A) Controls mounted on the side are brightness, focus, and astigmatism. (B) Power supply for the flying-spot scanner shown above.

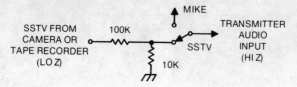

Fig. 3-9. Simple impedance-matching network for input/output circuits of major components in an SSTV setup.

Tape Recorders

As previously mentioned, an audio tape recorder quickly becomes an integral part of the SSTV setup due to the flexibility it affords. Although the tape recorder needn't be overly elaborate, good speed regulation and capstan drive are mandatory. Frequency response is not critical since the highest frequency handled is 2300 Hz; however, poor speed regulation will cause jiggle, which although not too objectionable, may be overcome completely by using a reel-to-reel machine. A reel-to-reel machine operating at 7½ IPS is superb; however, it quickly runs through all available tape. Thus 3¾ IPS is the best overall choice. The decision of whether reel to reel or cassette is mainly personal. Possibly the two-recorder setups would use one of each. A cassette recorder with fresh batteries or AC supply provides the most flexibility. A single program for each cassette cartridge may be recorded on both sides (directions) of the cartridge so that rewinding after use is automatic. These cartridges can be racked into a lazy Susan, similar to methods used in broadcast stations. End labeling of cartridges completes the setup.

TAPING PROCEDURES

When taping pictures off the air, care must be taken to insure on-frequency reception or pictures will not be reproduced properly. This can usually be accomplished on SSB by tuning for most natural sounding audio before starting the tape recorder. Later, two or three of the best pictures received during a QSO can be spliced or dubbed onto a QSO-made tape for later playback through the station monitor. Soon you may find you have a library of tapes and realize more than two or three pictures from each QSO defeats the basic purpose of a fast-moving reproduction that holds viewers glued to the screen. Presently 73 is the only magazine accepting tapes of SSTV pictures as "instant QSL signals" for SSTV awards;

however, it's apparent others may adopt this format in the future.

If taping pictures off the camera for later transmissions, several considerations will prove worthwhile. A speaker paralleled with your camera's audio output, as mentioned previously, for initial setup of the SSTV station will insure the proper video swing as it is recorded. Assuming all is operating properly, pictures now played back off the recorder should appear the same as those from the camera. If pictures look good on the monitor while being recorded, yet show dots or snow in the picture on playback from the recorder, increasing the recorder's input gain will overcome this poor video-to-noise ratio. If using the simple resistor-attenuator network (Fig. 3-9) described previously, lowering resistance values slightly will increase this level. After recording several pictures for a program, excessive repeats and blank spaces may be spliced out to give a continuously running tape. Be sure to record plenty of IDs for later splicing onto other programs or for tape loops. When splicing together programs, be sure to include the vertical-reset pulse before each frame. Leader tape added before and after each program helps later identification also. You can write on paper leader with a ball-point pen. If a program is recorded in both directions (sides), rewinding will not be necessary after each playing. Be sure to indicate the tape recorder input level so all frames will have the same volume when played back.

THE SSTV STUDIO

Although few amateurs have the room for a full studio type layout, careful planning and utilization of existing resources can result in a very professional setup. Keeping these thoughts in mind, the following ideas are presented.

Subjects for SSTV transmissions usually encompass an area of three feet or less. This reduces the large area and extremely bright-lighting conditions that are necessary in commercial studios. Many amateur operators obtain good results with only a single light bar (commercially available) mounted on the SSTV camera's tripod. This is a very flexible setup that allows the camera to be placed anywhere that is convenient; it can then televise anything appearing in front of the camera. This method uses *key* lights flooding the subject,

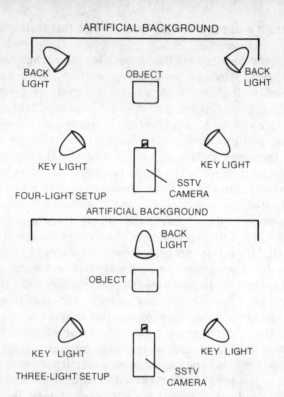

Fig. 3-10. Lighting arrangement of a typical home-brew SSTV studio. Either the three-light or the four-light setup benefits picture contrast. In either case, it is the addition of the back lights that aids picture quality. Use separate light dimmers on each light.

with *no* back illumination. While the amateur can obtain fairly good results with this on-light-bar/camera setup, two or three light bars *properly placed* in a suitable background will provide substantially better results; in fact, these results are often quite breathtaking. Proper placement of these lights will be discussed shortly. The additional lights have two distinct advantages: first, the f/stop is permitted to be closed down somewhat, allowing for a greater depth of field; second, this brighter picture produces a larger video level for amplifier stages, thus increasing the overall video-to-noise ratio. Individual light dimmers on each set of lights may then be used for controlling picture light levels. You may quickly find that a light dimmer is much easier to manipulate than a camera lens. SSTV differs from commercial TV in that 99% of the time specific objects are the main attraction, and a background is neither desired nor necessary. This fact immensely simplifies studio requirements and reduces the room necessary for operations.

A very efficient home-brew studio can be constructed using several sheets of ordinary black poster paper to form a wall or background. The SSTV camera is then positioned to look into this setup as in Fig. 3-10. The black background produces a dramatic effect in SSTV—not only does it make light-color subject material stand out in the picture, but it also permits all available light to show on the subject. A *key* light is placed on each side of the camera and one or two back lights are situated above and behind the objects being televised for illuminating the back of the object—not the background. The background is of minor importance in SSTV—back light (not background light) is very important to proper lighting balance in SSTV. It produces good contrast pictures with the illusion of a third dimension. Although few SSTV operators presently use back lighting, it should not be overlooked. The results obtained are much better than that of front flood only.

Commercial TV broadcasters use a large amount of back light on dramatic programming like personal interviews and soap operas. Often the ratio of back light to front light is 7:1: of course this ratio must be varied according to the subject material and the mood desired. TV stations employ lighting engineers for this purpose.

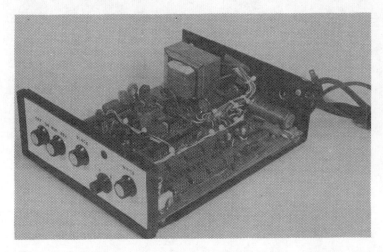

Fig. 3-11. A home-brew gray-scale generator is ideal for setting up the SSTV station. (Courtesy WA9MFF)

The SSTV camera should be capable of panning from left to right of the setup, plus move in various directions with minimum effort. One of the best methods of accomplishing this is to mount the camera tripod onto a wooden platform supported by metal rollers.

An SSTV extra that will help contrast and resolution is the gray-scale generator shown in Fig. 3-11 from WA9MFF. Of course this isn't a project that the newcomer might try on a weekend, but it has many obvious advantages over gray-scale tapes.

Chapter 4
Operating Procedures

As with any new mode of communication, it is suggested one takes a while before actually making transmissions to study other SSTV stations on the air. Listen to some of the older SSTV operators and notice how they break often to ascertain their SSTV pictures are being properly received. Notice how they shift frequency to avoid interfering with each other. How many pictures of the same thing does it take before you become bored? Notice how annoying *video only* ID signals are under poor copy conditions and how a station that is tuning up affects SSTV transmissions. Armed with a little foreknowledge, you will slip smoothly into the ranks of superb SSTV operators.

During the early days of SSTV (sixties), gathering frequencies were chosen so all SSTV operators could meet. Often (and even today) SSTV operators would leave their rigs "propped" on these frequencies so when activity started all could jump in and get pictures rolling. SSTV transmissions are no longer on just these frequencies but often extend 10–20 kHz higher in the band. SSTV operators remember that SSTV transmissions are legal anywhere in the Advanced class portions of 80–15 meters and all General class allocations are 10 meters. It is by "gentlemen's agreement" SSTV operators don't carry on long-winded picture exchanges with nearby stations in the first 25 kHz of any band.

Probably the most popular SSTV frequency used is 14.230 MHz, which is like a constant-running SSTV hamfest. There's usually an SSTV round table in progress that permits all involved to view quite an array of pictures for each individual picture transmission. The information and assistance exchanged on 14.230 MHz is quite phenomenal, and newcomers are always welcomed. Should you not care for this "round table" situation, rest assured your SSTV CQ signals on 14.235 or 14.238 MHz, for example, will be heard by other SSTV operators. As has been said so many times, shifting slightly up the band to avoid QRM is legal.

The suggested method of calling CQ is on audio with an SSTV picture or two interspersed every so often. This way your pictures may be noticed if perchance your audio isn't. Further, this will let non-SSTV operators know who you are and what you're doing, possibly eliminating being QRMed as an intruder in the ham bands. Remember some people know very little about SSTV or have misconceptions. You must take time and explain it to them. Keep in mind that we are modern-day pioneers in one of the newest frontiers of communications. It is up to us to present a pleasant, interesting image to others.

When you transmit complex SSTV pictures, try to explain them briefly beforehand so others will have an idea of what to expect. Remember to check often to ascertain your pictures are being received, and alter your number of pictures according to band conditions. Audio breaks during SSTV transmissions also give the rig a breather, which is advantageous when you consider SSTV demands are the same as CW key-down: 100% duty cycle. Watch the final amplifier tubes, and reduce power to half! A good procedure is to begin and end each transmission with audio rather than SSTV. This prevents leaving the other fellow hanging, wondering if it's his turn to transmit. Tell him verbally. One word here on *ethics*—the acceptance of SSTV today depends on how well we use this fascinating new mode of communication; just as distasteful language is undesirable on audio, distasteful pictures are even more undesirable on SSTV. Remember *you* help establish SSTV criteria. Scrutinize your pictures before transmission.

There are several SSTV nets presently in operation, where you can enjoy seeing and exchanging pictures from all

Table 4-1. Existing SSTV Nets

NET	VARIOUS NET CONTROL STATIONS	MEETING TIMES	MEETING FREQUENCY	CHECK IN PROCEDURE
U.S.A. SSTV NET	W9NTP K2KEY W1JKF W0LMD	1800 Z* Saturdays	14.230 MHz	Call in on audio—each caller in turn will then be called on to transmit SSTV
OCEANIA SSTV NET	VK3LM VK5BS	0100 Z* Sundays	14.230 MHz	Prefer initial call-in on audio—stand by for instructions
CANADIAN SSTV NET	VE7JA plus various VEs	2130 Z* Sundays	14.180 MHz	Listen for call area, then transmit short call on audio and await instructions.

*—Z indicates Coordinated Universal Time—what was called GMT.

directions and get assistance from other SSTV operators with any problems you might encounter. Table 4-1 is a compilation of the more active nets.

PRODUCING PROGRAMS

The best arrangement for operating convenience is to make up a group of program tapes for later transmission at will. This setup leaves your camera free for live transmissions while also permitting effortless, enjoyable QSO signals. It is much easier to sit back and watch your own pictures being transmitted than to always work with the camera. A simple SSTV/mike switch installed on your microphone soon proves quite helpful for video switching. Among the pictures you will want to make up for SSTV programs are pictures of yourself, your rig, XYL (wife), projects and ideas, special effects, plus plenty of different and catchy ID signals. If you have some ideas worth discussing, why not make up some pictures for illustrating them to others? Remember, SSTV can expand many present boundaries. Planning and ingenuity pay off here also. Don't try to put all your programs on one tape; instead use separate tapes for each set of thoughts, recording the same program on both sides of each tape to eliminate rewinding after every use. It only takes a second or two to swap tapes while you flip back to audio and give a narration of ID. This definitely beats hunting through the complete tape for a particular picture.

An assortment of ID signals is preferable. Some may be simple with large white letters on a black background for DX or poor conditions. (A group of 2 second quarter-frame pictures is ideal for this.) Other ID signals may be made fancy with unusual effects—checkerboard, diagonal letters, zig-zag lines—for the "closed-circuit" QSO calls with old-time friends. Notice commercial TV ID transmissions, magazine logos and billboards; there's thousands of catchy ideas around for you to add to your own ideas. An SSTV setup should also include various SSTV CQ calls designed to entice others to see more of your pictures.

Usually, three frames of each picture in a program is sufficient for good copy without becoming boring. The choice of stopping and starting a tape between pictures or splicing the pictures together is personal preference. Just remember to listen for the vertical sync pulse each 8 seconds, and keep a record of the pictures in each program as they go together. Always use AC or fresh batteries in your tape recorder. Clean tape recorder heads often and periodically check sync and frequencies of SSTV gear. A further discussion of tape programs plus a brief comparison of cassettes versus reels was included in Chapter 2.

SSTV DX

DX stations are often noticeable on the air by their *faster than usual* sync pulses. United States stations' sync pulses (60 Hz mains) sound like an old-fashioned movie projector simply clicking along, whereas DX stations (50 Hz mains) sound like an old-fashioned movie projector running too fast. Likewise, DX SSTV pictures are slightly narrower than United States pictures when displayed on an SSTV monitor. Noticing these things whenever you hear or see a DX transmission soon makes them easily detectable on the air. If you're attempting DX contacts under adverse conditions, try light letters on a dark background and partial frames which display your call more often. After initial picture exchanges have been made, you can transmit more detailed pictures and make longer transmissions. A rig can be pushed to near its rated limit for a very short SSTV transmission: Maybe two 2-second pictures, but the same demands on full 8-second frames would be rather excessive. The question of whether an additional S unit or so is worth that expense is strictly personal. Remember, the absolute maximum SSTV power allowable is 900W (because at any particular instant of time SSTV is a tone—the same as

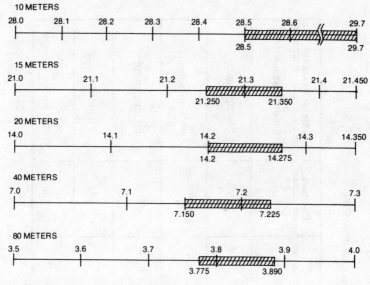

Table 4-2. Frequency Allocations of SSTV in Megahertz

CW) unless accurately measured, whereas 1000W input is the limit. But this absolute extreme is seldom necessary. If you consider that 500W is only 3 dB less than 1 kW (1 S unit), it hardly makes this legal-limit push worthwhile. There are presently around 95 countries on SSTV with more appearing daily. Many of these operators are the only SSTV station in their country. Your patience and understanding is vital to their success. Keeping this in mind may help you to become one of the first to make DXCC/SSTV.

POWER REQUIREMENTS

The SSTV newcomer must remember that SSTV is a 100% duty cycle—the same as CW—thus reduced power is definitely recommended. If conditions are good only 100W may be plenty, and if conditions are extremely poor even 1 kW may not help. Extensive tests have shown that an SSTV power level of half the average SSB audio output is near optimum. Thus, if a station is running 400W *average* output when talking, then the SSTV driving signal should be adjusted for an output of around 200W. Refer to Table 4-3. This usually results in an SSTV signal being the same volume level as the audio when received.

Table 4-3. Comparison of Power Levels for SSTV

Power Level	Picture Quality	Decibels	S-Meter Readings
50W	Some snow if band conditions are poor—SSTV picture is easily affected by any QRM	−6 dB	2 S units weaker
100W	Will produce good picture with very slight snow—signal will override light QRM	−3 dB	1 S unit weaker
200W	Usual SSTV power—this level will produce good pictures under minimum QRM conditions	Reference level	Reference level
400W	Solid copy SSTV pictures—signal will override all but the worst QRM	+3 dB	1 S unit stronger
800W	Solid copy picture—overrides QRM	+6 dB	2 S units stronger
1Kw	Same results as 800W	+7 dB	2.4 S units stronger

Remember, reduce the power when transmitting SSTV by lowering the SSTV level or transmitter input level (mike gain); not by detuning the final. Let the rig loaf when transmitting SSTV. If the final power amplifier tube plates show much color they are probably running too high an SSTV level. The difference between 400W and 800W, for example, is only one S unit, where each S unit is equal to 3 dB.

LIVE TRANSMISSIONS

Sooner or later you will want to try live SSTV transmissions. Whether the end result is worth the effort will depend on the forethought given. Find a location for the camera and lights that allows televising the operating position, but do not blank the monitor with brights lights. You must be able to view the monitor while the camera is viewing you. It is handy if the camera and lights can swing around 90° or 180° for viewing photos or slides placed on a wall. You should not need to rebuild an SSTV setup to televise printed material. Floodlights, SSTV switching, and taping should easily be controllable from the operating position. Marking focus points on the camera lens for live transmissions help tremendously. If for some reason you can't view your SSTV monitor when transmitting live, try connecting SSTV audio to an amplifier and speaker and listening for the vertical reset pulse each 8 seconds. This will prevent you from accidentally moving in the middle of a picture.

FRAMING SSTV PICTURES

Using the SSTV camera can be a very simple process: the operator merely directs it to the subject material, focuses, and transmits. If you have proper illumination and interesting subject material, the results are quite pleasing. If the subject material is printed matter previously laid out, this operation is quite simple; however, if the subject material is objects or people being directly televised then proper framing and picture composition are worthwhile considerations. Layout of these live pictures is an art in itself, limited only by your desires and patience. The most professional appearing pictures are those with proper balance and illumination. This balance situation varies with various subject matter, but it is generally reflected as pictures that are comfortable to view. For example, if there is one object in the scene being televised

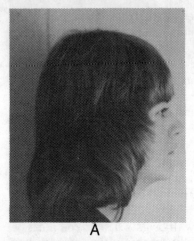 

Fig. 4-1. Framing pictures. (A) The face is centered in this picture, but an unbalanced effect is produced. The viewer is left wondering if there is a secondary object not shown in the transmission. (B) This shot is effectively balanced. There is enough nose and head room so that the viewer can see that there is not another secondary object. These photos are of the XYL of WA4APM/5.

it is placed in the center. If several objects are being televised, they are situated so both sides of the picture have equal content. This is much more pleasing than having several items on only one side of a picture. Balance is just having material equally placed for a smooth view, rather than an unequal number of objects on sides of a picture. Balancing parameters vary between subject matter. In fact, commercial studios employ production engineers for dealing specifically with this situation; however, the SSTV operator soon develops a good knack for pictures himself!

Televising people is the most obvious exception to the balance rule. Common reasoning gives us the explanation. When viewing a person for the first time, most people look at the eyes first, then the rest of the body second (Fig. 4-1). If a profile or sideview of a face is shown (and this is centered on the screen) the eyes will be near a picture edge. This gives the viewer a curious feeling because he cannot see what the televised person is viewing. In this case proper framing is obtained by panning the camera so the televised eyes are in the screen center. The viewer can then see if there is an object of secondary interest. In other words, watch about placing a nose in the picture corner. Leave some nose and top-of-head room so the viewer will feel like he is seeing all of the picture.

Slide projectors, especially those with remote controls, are perfect for the SSTV setup. All one needs to do is point the camera at a screen and focus. The projector provides its own light source, and you can select slides to be transmitted from your armchair.

OPERATING SSTV WITHOUT A MONITOR

Some DX stations use prerecorded tapes to transmit SSTV pictures while recording incoming pictures on another tape. Later they forward these received-pictures tapes to SSTV friends who display these pictures on their monitor, photograph them, and return the photos to the DX station. Sometimes SSTV operators carry tapes in their car to exchange pictures while mobile. The received pictures are then viewed when they return to the home station. SSTV has unlimited possibilities—picnic portable, horseback mobile, and snowmobile mobile are just a few of the fun ideas.

SSTV PICTURE ANALYSIS

While the majority of SSTV contacts may be solid copy with closed-circuit quality pictures on both ends of a QSO, there are times when the many variables of amateur radio can affect an SSTV picture received over the air. For example, multipath propagation can cause ghosts and loss of sync; or variations in tape recorder speeds can cause apparent skewing of lines in pictures. These mild variations are a fair tradeoff for such a narrow bandwidth long-distance communications medium. In order to give the SSTV enthusiast a fair insight into these idiosyncrasies, the more common occurrences will now be discussed.

Multipath propagation is caused by signals being received via two paths of different distances. An example of this is the ring or echo sometimes heard on DX stations (multipath propagation is most common on long-distance contacts). Let's assume the transmitting station is located in England. While one path of this signal travels directly over the Atlantic Ocean to the United States based receiving station, another path of this same signal travels over the Indian Ocean, across Australia, and then transversing the Pacific Ocean to the receiving station. Although both signals are traveling at the same speed, 186,000 MPS, one signal is delayed slightly (approximately 5 μs) because of the extra distance it

Fig. 4-2. Multipath SSTV signal. The ghost is to the right of each letter. Being very slight, this ghost indicates a low-amplitude delay signal and a strong initial signal. The salt-and-pepper effect in the picture is caused by noise.

transversed. When the two signals are received and reproduced, the result is one of the sounds being slightly delayed, producing an echo. This same situation holds true in SSTV except the results are seen on a monitor screen rather than being heard on a loudspeaker (Fig. 4-2). Each scanning line of an SSTV picture is 66 ms long, regardless of screen size. If a dark object or bar is being transmitted and multipath is experienced, there will be a *second* bar displaced by approximately 5 ms to the right of the initial bar. See Fig. 4-3. Should this situation hold true for the full 120 lines that ghost or shadow is produced on the monitor screen. The intensity of this ghost will depend on signal strength of the long-distance signal. If the long-distance signal is strong, the ghost will be quite noticeable; whereas if the signal is weak the ghost will be dim. The unusual phenomenon of multipath effect on SSTV is that the video frequencies (1500—2300 Hz) may be affected but not the sync frequency (1200 Hz), or vice-versa. The previously mentioned situation exemplified additive, or in-phase, multipath; however, there are also times when the multipath signals are subtractive, or out of phase. Multipath propagation is exclusively dependent on ionospheric conditions, so there is no way of predicting this in- or out-of-phase relation. Out-of-phase multipath signals cancel each other. This cancellation is most detrimental when it occurs at sync frequency. The most obvious example of this in SSTV is when the signal sounds perfect but synchronization cannot be achieved. This is because the out-of-phase sync pulses are cancelling each other. The net result is no picture displayed and random scanning.

Jiggled lines in SSTV pictures are caused by poor speed regulation of tape recorders. Although these minute variations are not apparent in audio applications, they are quite noticeable when sync pulses for several lines vary by 1 or 2 μs. Refer to Fig. 4-4. In other words, several sync pulses (which initiate each scanning line) may be delayed, thus displacing the respective video information.

Skewed lines superimposed on SSTV pictures are an entirely different situation. These white lines are caused by variations in tape speeds between record and playback modes (Fig. 4-3). Close inspection of the top lines in an SSTV picture will reveal four slight humps, or waves, in the picture. These waves are caused by a minute amount of 60 Hz AC hum influencing the horizontal scanning of a SSTV picture. If a constant tape speed is maintained, these waves are not noticed because all waves in the scan lines align. If separate tape recorders (or the same tape recorder under various states of battery charge) are used for record and playback, this slight displacement of "hum waves" can cause skewed lines. Refer to Fig. 4-5. These lines will lean from picture top left to bottom right if playback mode is slower than the record mode, and from picture top right to bottom left if playback mode is faster than the record mode. These skewed lines are most

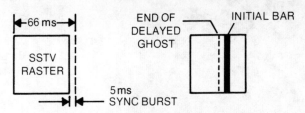

Fig. 4-3. How ghosts are produced. The illustration to the right shows the normal SSTV raster with its 5 ms sync burst. If the multipath signal is delayed enough, the half-frame ghost effect may appear.

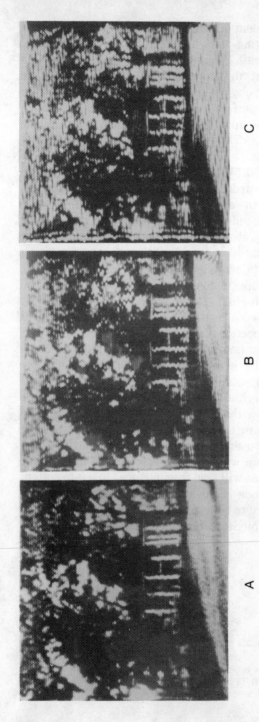

Fig. 4-4. SSTV picture of Arlington, a deep-South memorial home. (A) This transmission indicates perfect tape recorder speed. Notice the straight columns on the front of the house and the neat left side of the picture; however, the horizontal scanning lines have four minute waves. (B) This transmission has jiggled lines due to poor speed regulation of the tape recorder. Each horizontal line starts at a different location. Notice the jiggled columns on the front of the house and the jiggled left edge of the picture. Remember that this is typical of irregular tape speed. (C) This picture has skewed lines running from top left to bottom right. Skewed lines are caused by slow playback speed. Skewing in this transmission is created when each line of the four minute waves are displaced during playback.

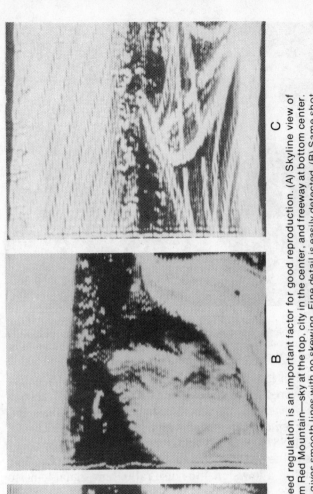

Fig. 4-5. Tape recorder speed regulation is an important factor for good reproduction. (A) Skyline view of Birmingham, Alabama from Red Mountain—sky at the top, city in the center, and freeway at bottom center. Ideal tape recorder speed gives smooth lines with no skewing. Fine detail is easily detected. (B) Same shot as above but the playback speed is slightly reduced. Notice that a few lines begin to lean to the right. (C) SSTV picture of the freeway system with Birmingham, Alabama in the background. Playback speed of the tape recorder is much slower than the recorded speed, resulting in extreme skewing. Attention should be given to the fact that the playback speed is slow but constant, causing skewing—not jiggle.

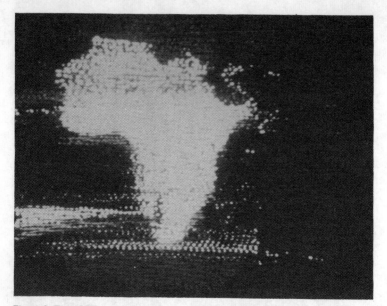

Fig. 4-6. This SSTV picture was transmitted from 2S3B in southwest Africa to K4TWJ in Alabama. Band conditions were poor, causing heavy noise in the video; however the sync pulses were not destroyed. Video noise is characterized by the salt-and-pepper effect.

pronounced in low-speed tape recorders (like cassette units) that are powered by batteries. Note that with skewed lines the tape runs *slower* but at a *constant* speed. Jiggled lines are caused by *irregular* tape speed within each 8-second picture.

Noise in SSTV video is apparent by a random display of white dots in the picture. This *salt-and-pepper* effect (Fig. 4-6) is the result of a low signal-to-noise ratio. Low signal-to-noise ratio could be due to a low-level signal feeding the SSTV monitor or a weak signal received over the air.

Partial loss of horizontal sync may be due to adjacent-channel interference, weak signals, or static. The end result is parts of the picture being lost or torn. Monitors utilizing PLL (phase-locked-loop) circuits in horizontal sweep will tear, locking to the picture when sync pulses are again received. Monitors with triggered sweep require a sync pulse for each line so that no video information is produced (black space in raster) until sync pulses are received. See Figs. 4-7 and 4-8. The choice of which system a monitor uses is strictly academic. Each one has certain advantages and disadvantages. A further discussion of these systems is given in Chapter 5.

A

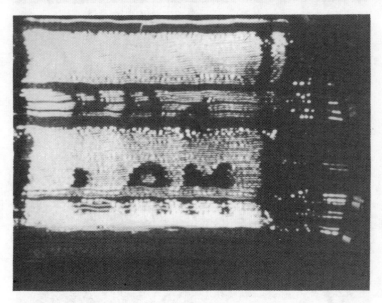

B

Fig. 4-7. Horizontal sync problems. (A) Loss of horizontal sync on a triggered sweep monitor. Lines were lost due to the lack of initiation pulse. (B) This is the same transmission as above but the loss of horizontal sync appears at a different location in the picture. This was transmitted from PA0LAM to K4TWJ under poor band conditions. The signal strength was approximately S5.4.

A

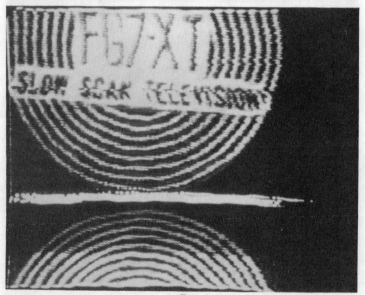

B

Fig. 4-8. Vertical sync problems. (A) This split picture was caused by loss of the vertical reset pulse. The monitor was reset manually. During each vertical retrace, the pulse is 30–50 ms at 1200 Hz. (B) This split picture retraced automatically when the scan line reached the bottom of the screen. The vertical sync pulse was not received at all. Both of these transmissions were transmitted from FG7XT to K4TWJ.

Chapter 5

SSTV Monitor Circuits

The most important and necessary part of an SSTV setup is the monitor. Due consideration should be given toward securing the most outstanding monitor feasible for your particular case. Features to look for in a monitor are reliability, quality, type of interference protection, and simplicity of design whenever possible. This may seem like a lot to ask of a unit, but not necessarily if we are willing to accept a small compromise in each of these areas. In order to understand this, let's take a closer look at each of these parameters.

Reliability and quality are very closely related in a monitor. Speaking from the standpoint of commercial units, these terms describe a circuit that has proven itself over a reasonable period of testing or operation; one that used quality parts, no more than necessary for the overall performance. If a monitor is being home brewed, namebrand parts quickly tested before installation and glass-epoxy printed circuit boards assure initial and long lasting success.

Interference protection determines how well a monitor works under adverse conditions. As mentioned in Chapter 2, most monitors perform equally under solid-copy conditions, but only the more sophisticated units excel under marginal conditions. The most common cause of this poor copy situation is adjacent-channel interference to sync pulses.

Before continuing this discussion, let's consider exactly what comprises an SSTV signal—1500−2300 Hz variations

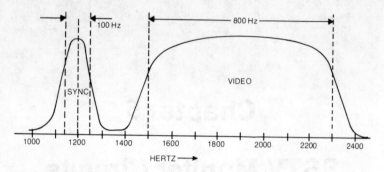

Fig. 5-1. Bandpass frequency spectrum of SSTV. The sync bandpass curve is 100 Hz wide and the video bandpass curve is 800 Hz wide.

(produce light and dark areas of the picture) and 1200 Hz sync pulses (short burst for starting each of the 120 scanning lines and a long burst for starting each picture, or each vertical scan). Should interference be present somewhere in the 1500—2300 Hz region, there will be a slight interference to the picture; however, picture reception is still possible. Notice that this full 1500—2300 Hz frequency range must be received for good picture reproduction. Not only do these video circuits need a steep sided response curve but they also require a wide bandwidth, compared to sync information. This is illustrated in Fig. 5-1. The fact that sync circuits require only a 100 Hz bandwidth is also shown. Sync pulses are important because they determine exactly where the video information will be placed in a picture. Should interference be encountered on the sync frequency of 1200 Hz, the picture information will be completely lost. When considering bandwidth and information content, it is more desirable to incorporate interference rejection into the sync circuits than in video circuits. Utilizing this information, let's now briefly discuss the sophisticated circuitry necessary for interference protection.

The simplest way of securing protection from adjacent-channel interference to sync is by using toroids or active filters sharply tuned to 1200 Hz. This virtually excludes any signal other than the desired sync signal from triggering the monitor sweep.

Other interference problems evolve from random noise overriding weak sync pulses or accidentally triggering the sync circuits. There are two popular methods of eliminating these problems and both methods presently use IC chips. One interference-reduction method uses gating and precise timing. The gates open only when sync pulses are to be received, close when sync pulses are not to be received, and close while each line of the picture is being scanned. This method eliminates false triggering during scans. If sync pulses are not received at the proper reset time, then those particular lines of the picture are omitted. When sync pulses again appear, the gates pass them, and normal scanning resumes. This method is often called triggered scanning, or triggered sweep. The other method of eliminating interference to sync utilizes a PLL (phase-locked loop). The PLL continually produces sync driving pulses near the desired rate (15 Hz for horizontal scanning). These driving pulses are used to trigger the horizontal sweep circuit. Incoming sync pulses serve to lock the PLL output pulses to the proper rate. In addition, the PLL will continue to deliver output pulses at the exact rate of the incoming SSTV should a few of the individual line sync pulses be missed. During these no-sync-pulse times, this method produces white rather than black on the monitor screen.

Simplicity of design means accomplishing all of the desired monitor functions in a straightforward manner and with a minimum of parts. Not only does lavish use of parts make a monitor unnecessarily complicated it also increases the possibility of breakdowns. Just as four resistors have more chances of failure than two resistors, a monitor with 30 IC chips has more possibilities of developing problems than a monitor with 15 IC chips. This, however, in no way implies that simple monitors are preferred over the more sophisticated monitors. Indeed, the sophisticated monitor is vital for *solid* communication on the crowded amateur bands of today. What simplicity of design does mean is that a monitor should have a definite need and use for each component. If an LC combination (passive) will perform equal to an active filter using an IC, the LC network may be the more logical choice because it will develop less trouble during operation. Should the front end of a monitor use two or more transistors for amplification and limiting while these same functions could be handled by one IC, the IC is preferred. Because modern technology considers an IC as a discrete component—the same as a single transistor—one component will give less trouble than two or more components. Interference protection circuits are paramount and sophisticated circuitry is a necessity;

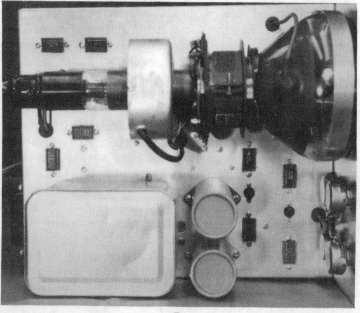

Fig. 5-2. The W9LUO monitor. (A) Front view of the monitor with all operator controls readily accessible. (B) The top view of the monitor shows the TV-type yoke used for deflection, focus coil assembly, and plug-in IC chips.

however, an abundance of unnecessary elements in these sophisticated circuits can be problem sources. Even technical engineers refer to exotic circuits with the slogan of "keep it as simple as possible for the specific job."

The previous standards were presented as a guideline for those individuals desiring evaluation criteria, but this is far removed from an outstanding consideration—your personal preference. One monitor may have a specific feature that makes it more appealing than all the others to you. Our hobby is meant to be enjoyable, and it's a fact that you receive more actual enjoyment from something that really strikes your fancy. Fortunately, amateur radio is one area in which these whims are affordable.

W9LUO MARK II MONITOR

Several years ago, Bob Tschannen, W9LUO, of Lombard, Illinois designed the basic circuit of this monitor (Fig. 5-2). Soon thereafter he began modifying it for even better results. Bob felt the sync circuits could be improved and video modulation to the CRT could be increased. The result is a truly outstanding unit, bearing only a slight resemblance to the original monitor. This unit reflects good design techniques using inexpensive components which are readily available. The circuit was proven through many hours of on-the-air use. Creditable operation under interference conditions is afforded by sharply tuned circuits and the horizontal PLL circuit. The simplicity of design makes this monitor (Fig. 5-3) ideally suited to the SSTV newcomer.

Circuit Description

Several connections are shown on the monitor's input (J1, J2, J3) for the various video sources in a typical SSTV setup. From here, signals enter the monitor's first stage, a 741 operational amplifier. This stage provides a substantial amount of gain with some limiting. The circuit will limit symmetrically with about 0.25V P-P input. As a typical ham receiver will provide several volts of drive, this necessary 0.25V level is quite easily obtained at low speaker volume. In other words, U1 merely amplifies and clips the incoming SSTV signal. Immediately following the amplifier/limiter is a discriminator (quadrature detection) consisting of L1 and C2. This circuit is tuned to 2300 Hz, and its function is to extract the

FM subcarrier from the slow-scan signal, slope detect it, and develop the resulting AM signal across C1. At this point, the signal splits in two directions, both through 1K resistors. One signal goes to L2, which is tuned to 1200 Hz. This signal then drives U2. The CA3046, U2, is a transistor array consisting of five silicon transistors: two of these are connected in a common-emitter configuration. Output of the push-push sync separator is taken from pins 1 and 5 of U2. Because a full-wave sync detector is used, the output frequency is doubled, making it much easier to filter. Some gain is also obtained in this stage.

Horizontal sync drive from U2 is now coupled through the 1 μF capacitor to U3, the NE565. This PLL circuit is used in order to provide better noise immunity than is obtainable from normal driven sync systems. The reason for this is that the PLL will free run regardless of input pulses and provide a sweep-trigger signal, whereas with a *triggered-sweep* system those scanning lines would be omitted. A string of square-wave pulses emerge from pins 4 and 5 of U3, the PLL. These are processed through an amplifier in U2 and output on pin 11. The pulses are differentiated by a 0.1 μF capacitor and applied to the base of Q2. Each time a positive pulse hits the base of Q2, the transistor conducts and discharges the 2 μF capacitor between its collector and emitter. This forms a sawtooth wave that drives U4, producing horizontal-deflection drive, which then powers complementary transistors Q3 and Q4. Picture centering is accomplished by adjusting the DC offset voltage of U4. This varies the deflection current through the yoke by varying the crossover point for the complementary transistors.

During this same time period, sync pulses from pins 1 and 5 of U2 are also applied to the vertical integrator. This integrator is directly above U2 in the schematic. After integration through the 4.7K resistors and associated 10 μF capacitors, this pulse is amplified and clipped in U2. The pulse leaves on pin 8 or U2, and then goes directly to pin 4 of U5, the vertical monostable multivibrator. A monostable multivibrator is preferred over a PLL in the vertical section because of the slow-scanning rates. The lockup and acquisition time would be too long for a PLL. Three or four frames would be needed for it to lock on to incoming SSTV; thus, driven sweep is employed in the vertical section. The vertical-retrigger pushbutton is provided in case a vertical pulse is missed. Output pulses from U5 are used to drive Q5 into conduction at the start of each picture. This discharges the 8 μF capacitor between collector and emitter of Q5, producing a sawtooth wave to drive U6. This 72741 is an operational amplifier powered by +15V and −15V for linear amplification of the sawtooth. Picture centering is accomplished by varying the DC offset voltage applied to the operational amplifier. Outputs from U6 then drive the vertical complementary transistors, producing deflection. Under normal operating conditions, there will be one vertical sawtooth produced for 120 horizontal sawtooths. Special attention is directed to the 8 μF vertical sawtooth capacitor (**) and the 2 μF horizontal sawtooth capacitor (*). These are tantalum capacitors with very low leakage. These high-quality capacitors are necessary because they determine sawtooth linearity and picture shape.

Now let's return to discriminator L1 and C2 to follow the video path from C1. This path goes through the other 1K resistor and to the input side of transformer T1. The full-wave video detector is on the secondary side of T1. Full-wave detection is used because it will extract subcarrier information at twice the initial frequency. This system gives maximum frequency separation and allows filtering with smaller components. In other words, the FM subcarrier is converted to an AM carrier at L1 and C2, then a conventional full-wave AM detector extracts the useful video information. This video information is then passed through a low-pass filter that consists of L3 and C4. An 88 μH toroid (L3) is used in this stage because it is readily available surplus. The unwanted subcarrier is now discarded, leaving only video information. This video information is then amplified by Q1 and applied to the cathode of the CRT. The 2N3440, Q1 is a high-voltage transistor with a V_{CEO} breakdown of approximately 300V. This rating is quite sufficient since a supply voltage of only 100V is utilized. The video modulation applied to the CRT can be varied by the contrast control in the emitter of Q1. This video level is adequate for any popular P7 tube. The schematic shows a 5FP7, because they are readily available surplus.

This monitor applies video drive to the cathode rather than control grid of the CRT. This is done to eliminate degeneration in the cathode circuit. Cathode drive produces the brightest picture with minimum video drive.

There are several other CRT types that work very well in this monitor. The 5FP7 or 7BP7 work nicely when used with an

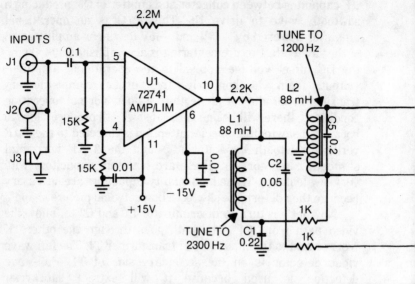
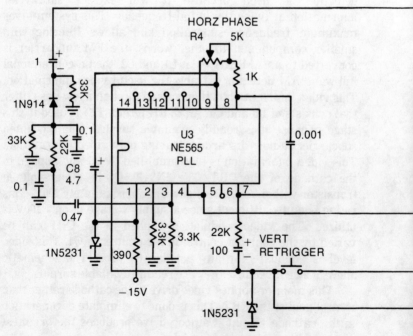

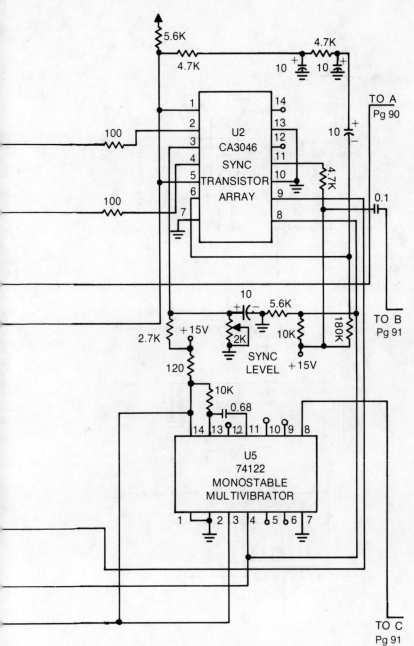

Fig. 5-3. Schematic diagram of the W9LUO Mark II monitor. All video and drive circuits are located in the left portion of the diagram. The circuitry employs six IC chips and seven transistors.

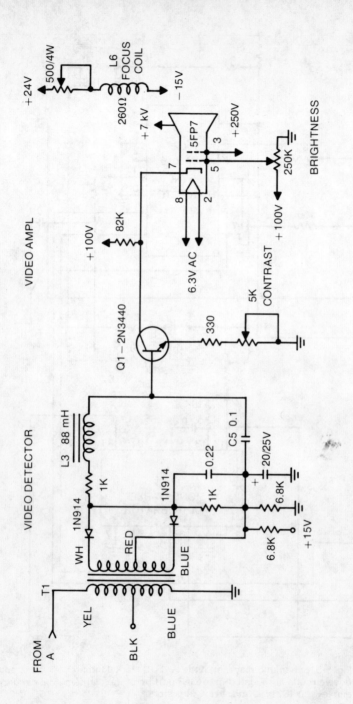

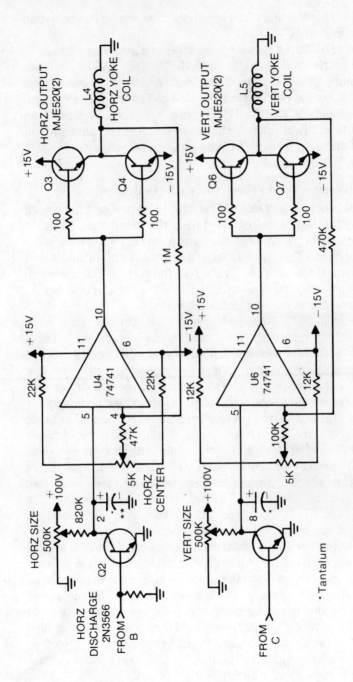

Fig. 5-3. (cont.)

old 50° TV yoke and focus assembly. Focus coils can be wound in many cases.

A 12BP7 produces large pictures and also uses a 50° or 70° yoke. An ideal tube is a 5AHP7. The 5AHP7 is electromagnetically deflected and electrostatically focused. This eliminates the need for a focus coil and considerable current. Unfortunately the 5AHP7 is not available surplus, only commercially at fairly high prices; however, its advantage of an aluminized screen produces very bright pictures.

Electrostatic Versus Electromagnetic Deflection

Another area of interest relative to P7-type monitor design is the method of deflection. Inside the monitor's CRT is an electron gun which fires electrons at the screen, producing a bright spot. This spot is then deflected both vertically and horizontally to produce a raster on the screen. There are two ways of accomplishing this deflection of the electron beam: electrostatically or electromagnetically.

Early SSTV monitors used CRT devices that were originally intended for oscilliscope operation. These electrostatically deflected and focused CRT types worked quite well in monitors for three reasons: first, the low accelerator voltage of 2—3 kV was easily obtainable; second, requirements for sawtooth waves of the voltage necessary for horizontal and vertical deflection plates were ideally suited to tube-type circuitry; third, focus voltage could be obtained from the deflection amplifier circuits. The tube-type monitor was a perfect match of parts and application. Transistors quickly altered this situation because they were not high-voltage devices. If a high voltage was applied to a transistor it was destroyed, but the transistor could handle a fair amount of current before it burned out.

Deflection yokes that require *current* for deflection (low impedance) have started to grow in popularity. Yokes were an ideal match to complementary output transistor configurations. Transistor oscillators used in pulse-type power supplies also began to grow in popularity. These units could use TV flyback transformers and easily produce the high accelerator voltage necessary for electromagnetic deflection. Thus, electromagnetically deflected CRT types and transistor circuits grew hand in hand toward popularity.

Power Supply Considerations

The outstanding point of interest in the high-voltage power supply (Fig. 5-4) is that it is a pulse-type unit made with a flyback transformer from a transistor fast-scan TV. Not every flyback can be used. For example, if you use a flyback out of a tube-type receiver, the circuit impedances will be much higher than those of the transistor flyback; therefore it will not function.

In the correspondence the W9LUO has received from hams, many have related the use of the wrong flyback transformer. The ones that have used the correct flyback transformer, but are still having problems, have usually used the wrong horizontal output transistor. This problem has been encountered in hundreds of applications on both the MK I and the MK II monitors. The correct types are the MJ105 (not MJE105) and the Motorola 48S134995. Many people relate that they are using a big transistor, usually 50W, but can't get any voltage out of it. The problem is that the turnoff time of any transistor, except TV deflection type, is much too long. So, one needs a fast-switching transistor for the power supply. Practically any output transistor that is used in the horizontal deflection circuitry of a solid-state receiver will work well.

One area of confusion is that transistor switching is only 15–16 kHz, and some believe that the speed of any other transistor is sufficient—this is a wrong assumption. In fact, using the wrong transistor, one will probably end up with only 2–3 kV. With the right transistor, the voltage will increase immediately to 7–8 kV, and the power supply will work smoothly.

You'll notice that three output voltages are available from the power supply. The output from the 100V zener regulator powers the discharge circuit and the video amplifier. The second anode of the monitor is powered by the output of the 7 kV lead, and the 350V supply drives the screen.

Monitor Tuneup and Alignment

We will now consider the tuning of L1 and C2 (Fig. 5-3). When this circuit is at resonance, the voltage across C1 will be minimum. This is critical and should be checked carefully with an AF signal generator and an oscilloscope. By swinging the input frequency to J1 back and forth between 2200 Hz and 2500 Hz, you can make sure that the dip does occur at

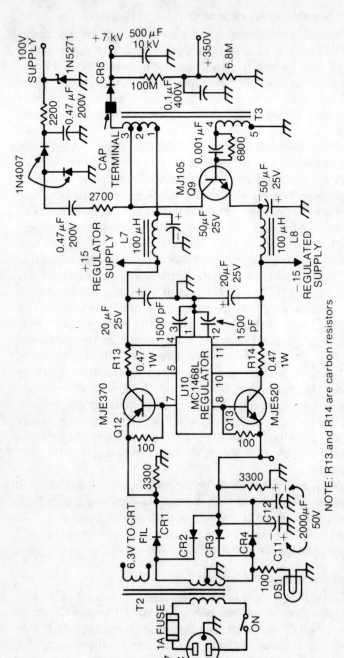

Fig. 5-4. Schematic diagram of the power supply for the W9LUO monitor. The power supply generates the high voltage for the CRT as well as the 350V, 100V, 24V, and ±15V supplies.

NOTE: R13 and R14 are carbon resistors

approximately 2300 Hz. This frequency can be varied to 2350 Hz or 2400 Hz if desired by changing the value of C2. A 5% change in frequency can be accomplished by altering the value of C2 by 10%.

Now move the scope input to either side of L2. Since this is acting as a balanced circuit, the voltage on either side will be essentially the same. You might wonder why both sides of L2 aren't driven. In any auto transformer there is very tight coupling; therefore, if one side is driven, an essentially equal and opposite voltage will be produced on the other side. This gives you two essentially equal voltages of opposite phase coming out of L2 to drive the push-push sync separator (U2).

After moving the scope input to L2, adjust the AF signal generator to 1200 Hz. Then adjust the value of tank capacitor C5 to give maximum output. A value of 0.2 μF $\pm$ 10% will normally give the correct resonant frequency.

After tuning these two circuits, the remaining tuning process consists of running an SSTV signal, preferably a tape, into the monitor. Adjust the sync levels so that the monitor is stripping sync properly. Then make sure that both the vertical and horizontal multivibrators are firing. To stabilize the picture, adjust the horizontal frequency with the phase control on the PLL.

W0LMD SSTV MONITOR

While the major interest among experienced SSTV operators has been in the *new* slow- to fast-scan converters, the beginner is interested in a simple trouble-free system with minimal cost. This SSTV monitor has been designed to satisfy a number of important yet somewhat conflicting objectives.

- Picture quality equal or superior to any home-brew or commercial monitor.
- A monitor whose basic operations are simple to use for the beginner, net operator, and SSTV contest operator.
- All-IC design with available low-cost parts.
- A high-quality PC board available at low cost.
- CRT and high-voltage power supply flexibility to permit using available surplus parts.

This W0LMD monitor (Figs. 5-5, 5-6, and 5-7) comes very close to satisfying all of the above objectives. The circuit uses

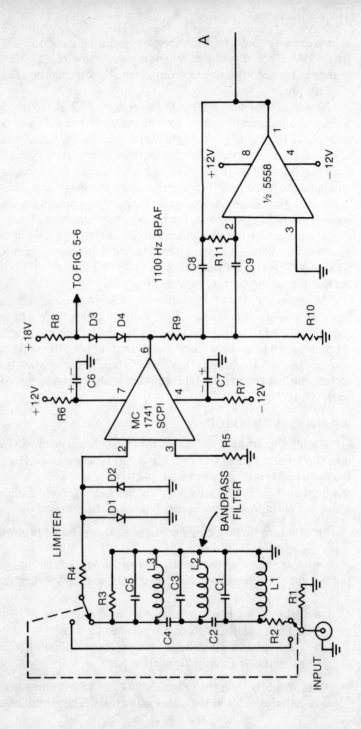

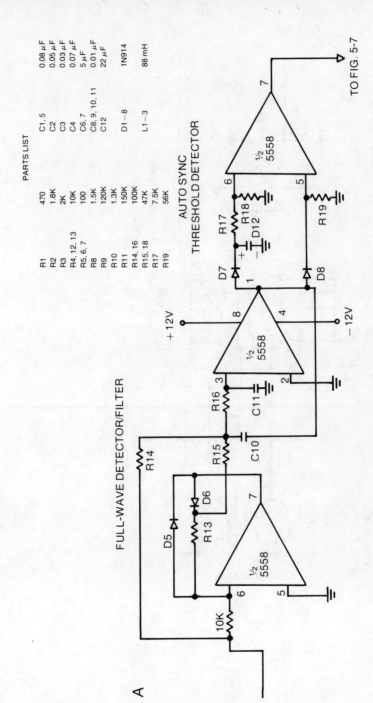

Fig. 5-5. Schematic diagram and parts list for the input and sync detector circuits in the W0LMD monitor.

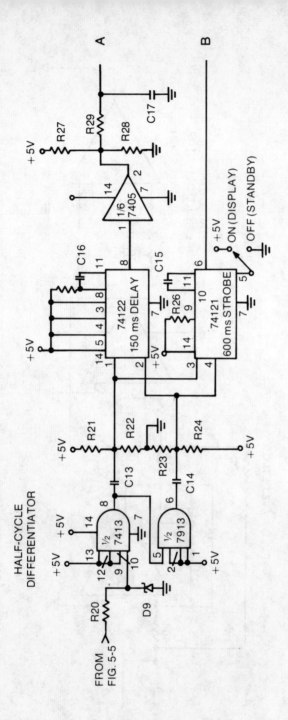

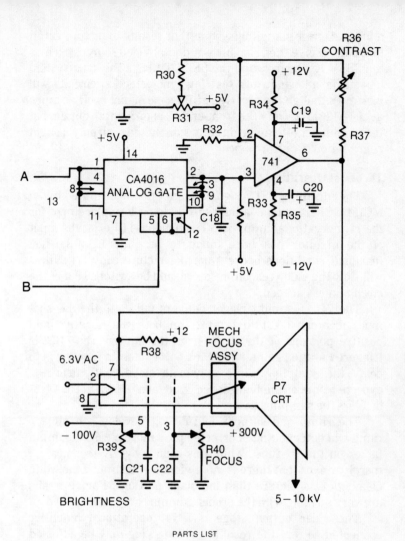

Fig. 5-6. Schematic diagram and parts list of the video circuits of the W0LMD monitor. Focus assembly is from surplus radar. Resistor R33 may be constructed from a string of 22M TV high-voltage resistors.

a video-carrier sampling system that results in picture detail that can resolve the crisp lines of digital video sources such as the SSTV Keyboard[1] and the SSTV Titler[2]. The sync system has sufficient bandwidth that the sync-detection circuits will lock even though the received SSTV sync signal is off as much as 200–1200 Hz. Only the front-end IC is difficult to obtain, but the ubiquitous 741 can be directly substituted with only a slight degradation in performance.

Theory of Operation

The incoming SSTV signal is amplified and limited by an MC1741SCP1, which is a high-speed Motorola replacement for the 741 operational amplifier. The bandpass filter at the input of the monitor has been found to be quite helpful when marginal conditions prevent normal picture copy. The filter will slow the video response, so it should be switched out of the circuit unless needed.

The 7413 Schmidt trigger delivers pulses to the one-shot multivibrators (7421 and 74122) on both the negative and positive portions of the signal from the limiter. The 74122 is triggered at the beginning of each half-cycle. Its function is to delay the sampling *ramp* (sawtooth slope) until near the expected time of the highest received tone, 2300 Hz. The 74121 delivers the sampling pulse.

The ramp and sample/hold circuit develop a nonlinear ramp and detect output, corrected for a more pleasing gamma factor on the P7 tube. While this sounds overly technical, it merely means that there is more video swing around the white elements of a picture than the black portions. Facial details are very striking with the proper gamma correction applied.

The video output stage is a 741 operational amplifier connected so that the resultant video swing may be adjusted via the *contrast* control. About 15V of video swing is adequate for the 7AUP7, but up to 30V of video swing may be obtained by connecting the 741 video output amplifier to +18V and −18V. Most people mount the *contrast* control on the front panel, but it may be mounted inside as a *set-and-forget* control. The *brightness* control is front mounted.

The sync circuits are a bit unusual. Sync detection is done by slope detection. An 1100 Hz bandpass active filter will cause

---

1. Suding. *CQ Magazine*. Sept. 1974, page 20.
2. Suding. unpublished design. SASE for details.

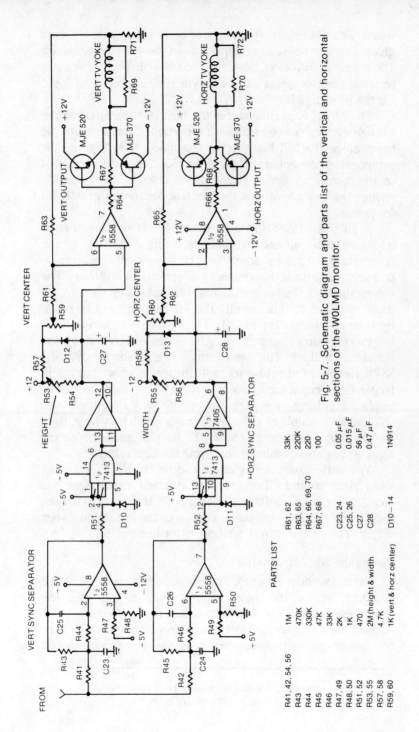

Fig. 5-7. Schematic diagram and parts list of the vertical and horizontal sections of the W0LMD monitor.

PARTS LIST

R41, 42, 54, 56	1M	R61, 62	33K
R43	470K	R63, 65	220K
R44	330K	R64, 66, 69, 70	220
R45	47K	R67, 68	100
R46	33K	C23, 24	0.05 µF
R47, 49	2K	C25, 26	0.015 µF
R48, 50	1K	C27	56 µF
R51, 52	470	C28	0.47 µF
R53, 55	2M (height and width)		
R57, 58	4.7K		
R59, 60	1K (vert & horz center)	D10–14	1N914

101

whatever is lowest in frequency to be labeled sync, which is given a larger boost in amplitude than the video frequencies. The output is full-wave detected and filtered. This results in a series of positive-going bumps at sync time as measured at pin 7 of the filtering 5558.

The 22 μF capacitor stores the high-level bumps, and the voltage-divider resistors pick off a portion of this as a sample base. Since the full bump level is applied at sync time, the comparator-connected section of the 5558 then picks off the top of the bump as sync. Midadjusted video sources merely change the bump level, not the bump action, unless greatly off frequency.

A pair of LPAF (low-pass active filter) devices provide additional filtering and threshold detecting of the vertical and horizontal sync pulses. Since the sync signals are considerably delayed, a standard fast-recovery sweep circuit is utilized. The values shown in the horizontal- and vertical-sweep sections are those required in this circuit. The 4Ω resistors can be made considerably larger (up to 22Ω) if less sweep is required.

Power supply design is left to the discretion of the individual builder. The power supply shown for the W0LMD SSTV Keyboard should work well for the monitor, but put in larger filter capacitors, say two 10000 μF or more to reduce ripple due to large sweep transients. The *SSTV Handbook*[3] and *QST*[4] show a couple of different circuits which may be utilized for the high voltage to the CRT. Surplus modular power supplies are also usable for supplying the high voltage.

The deflection yoke is usually some old 70° TV unit that was lying around. Remove any attached condensers and connect the high-resistance section to the vertical sweep output and the low-resistance section to the horizontal sweep output. The choice of the CRT is up to the builder.

Tuning the W0LMD Monitor

First, without the PC board inserted and the CRT attached, check the power-supply voltage. Be sure the *brightness* control is towards the −100V direction, then connect the CRT—be careful of any stored high voltage. Turn the power on and *slowly* bring up the *brightness* control

---

3. *SSTV Handbook*. 73 *Magazine*. page 89, 114.
4. Tschannen. *QST Magazine:* March 1971, page 37.

(towards 0) until a spot appears on the screen. Being careful to avoid burning the CRT, focus the dot and center it as much as possible if you have the recommended surplus mechanical radar assembly.

Turn the power off to insert the PC board. Tune in an SSTV signal. With the *contrast* near 0Ω (minimum contrast) adjust the *video symmetry* pot for rough symmetry around 0V as measured with a DC scope at *video out* (741 pin 6). Turn up the *contrast* a bit for about 20V P-P swing, and trim the *video symmetry*.

Tune up the sync by first checking for horizontal and vertical sync at pins 5 and 13 of the 7405. Tune the *width* pot for a slight overscan, adjusting *horizontal center* to center the horizontal sweep seen on the CRT. The emitters of the output transistors make a nice point to look at the resultant sweep. Avoid a flat portion of the sweep ramp. Tune the *vertical sweep* controls similar to the *horizontal* controls. Should you have backwards sweep, reverse the affected deflection coils.

RF FEEDBACK

One other area of interest concerning SSTV monitors involves RF feedback. This problem is characterized by interference to monitor-displayed pictures while transmitting SSTV. The amount of interference is usually directly related to the amount of transmitted power and antenna-radiation characteristics. As the SSTV monitor is a very sensitive unit, any amount of transmitted RF energy appearing at its input will cancel picture reproduction (Fig. 5-8). Thus, the solution to RF feedback is to eliminate any RF energy in the shack. The easiest method of RFI reduction involves connecting all equipment to a common ground. This ground should be *independent* of any other station wiring. Power cable grounds and mike shields are not sufficient for proper grounding. One method of creating a good ground is by placing a large copper strap behind the operating desk, connecting each piece of equipment directly to this strap, then connecting this strap to earth ground. One word of caution concerning grounding: Don't assume that because your rig is connected to an earth ground that it is actually grounded. For example, if a ground wire extends 16 feet before actually touching earth ground, it may be a 1/4-wave stub for 20 meters. If the ground wire does radiate, it will bring RF energy directly into the shack. For

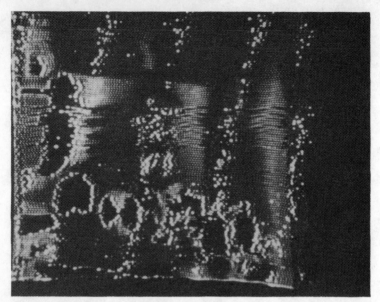

Fig. 5-8. Feedback as displayed on an SSTV monitor. This problem was caused by transmitted RF energy returning to the video stages of the receiver—all information is lost.

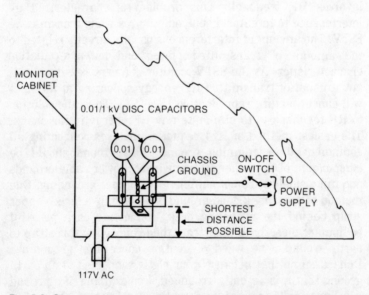

Fig. 5-9. One method used to eliminate RF feedback is to install bypass capacitors close to the 117V AC input to the monitor. Mount the capacitors as close as possible to the place where the AC enters the cabinet.

this reason, ground leads must be kept short as possible. The described grounding method will eliminate *all-but-the-worst* RF feedback problems.

Should some RF feedback still be evident on the SSTV monitor, both sides of the AC power line can be bypassed to ground with 0.01 μF capacitors as illustrated in Fig. 5-9. These bypass capacitors should be connected to the monitor AC input precisely where it enters the SSTV monitor cabinet.

If RF energy manifests itself in the AC power lines, it may also be necessary to bypass power cables on the SSB transmitter, linear amplifier, SSTV camera, and tape recorder. Another feedback problem that may arise is when the antenna brings a portion of the transmitted energy directly into the shack. Although these problems are most common with end-fed long-wire antennas connected directly to the transmitter, they can still develop when a dipole or beam antenna places the field of maximum radiation near the SSTV monitor. This RF saturation is very difficult to eliminate using conventional bypassing and grounding techniques. When RF is eliminated from one stage or area of a unit, it instantly appears in another stage. The most effective solution to the predicament usually involves separating the antenna from the shack. This is sort of a last-resort solution.

Chapter 6
Digital Scan Converters

Since 1958 when SSTV was born, technical advancements have been phenomenal. This is, no doubt, attributed to the large number of devoted technicians and engineers who enjoy the challenge of SSTV development. These individuals have accelerated SSTV technology at an exponential rate, while merely enjoying a great hobby.

The first SSTV units were bulky and used large components—vacuum tubes, electrostatically deflected CRTs, and high-voltage transformers of large-current ratings.

Next came transistors, and SSTV experimenters were among the first amateur operators to actively use these in video communications gear. During this same time, other SSTV operators began experimenting with electromagnetically deflected CRTs. On-the-air discussions of these activities served to accelerate advancements. Through joint efforts, a completely new state-of-the-art evolved for SSTV. Monitors became transistorized, using electromagnetic deflection and employing transistorized oscillator type high-voltage power supplies—progress was tremendous! As a particular unit was built and debugged, the designer or designers would outmode it and build a better unit. Evolution was endless as each SSTV experimenter worked diligently to improve existing units.

IC chips were the next evolution in electronics, and SSTV experimenters put them to work. Long before information on these chips was widely published, experimenters were using these devices to replace transistor functions like amplification, limiting, and frequency separation. Further discussion of IC chips now calls for separating them into two categories—linear devices and digital devices. Linear IC chips are operational and differential amplifiers that give *linear* reproductions of input signals. Digital IC chips can be used as gates, flip flops, and shift registers. These *logic* devices use two voltage levels for operations using the binary number system. As digital IC chips are one of the latest electronic innovations, this chapter is concerned with their applications in SSTV. It must be realized that digital electronics includes everything from MOS shift registers and random-access memories to charge-coupled devices. These are areas where advancements are presently making tremendous progress. With these thoughts in mind, the simple principles of converting TV to the digital world will now be shown.

ELEMENTARY PRINCIPLES OF DIGITAL TV

You can look down at your hands and be instantly reminded that the world counts in powers of 10. Nature provided us with 10 fingers and 10 toes, and until recently this method of counting was nearly universal. Our entire number system is based on the power of 10. After you run out of fingers, just move over one column of counted fingers, and do it over again. As an example: 10 is 10^1, spoken 10 to the first power; and 10^2 is 10 to the second power, or $10 \times 10 = 100$. This is carried out indefinitely as our need for larger and larger numbers grows. It should be noted that the larger the base is, the easier that large quantities of items can be counted.

All of this is fine, but is there a reason for going to smaller bases? Nature seems to like the binary system (base 2). It will probably be shown someday that the brain works on the binary principle. In any case, it turns out that electronics is easy if it just makes simple decisions of *on* or *off*. Amplifiers do not need to reproduce waveforms, just recognize an *on* or *off*. Sounds easy doesn't it?

Let's take a look just how you can live in a two-finger counting world: You might want to order six oranges over the phone. Since you must describe your desires to the ordering clerk without using the numbers 6, 5, 4, 3, and 2, you might be a bit perplexed. You are only allowed to use numbers 1 or 0. You can easily see how to count up to 2. The solution is to do exactly what you do when you count to 10 in the decimal system. You move over a column. The proper designation for the decimal 2 is *one, zero* (1,0) in the binary world. This is not signified by the word 10—just 1, 0—since people have not taken the time to name it as was done in the base-10 system. Before you count farther, you should be aware of the fact that the second digit is more significant than the first digit. This is also true in the base 10-system. Take the example of 12. The 1 means 10, and the 2 means add 2 to the 10. The far-left digit is called the most significant bit (MSB). The far-right digit is called the least significant bit (LSB). This designation is important and will be essential in the understanding of all digital counting schemes.

Let's continue on, the six oranges aren't ordered yet. In the binary system 2 is 10, 3 is 11, 4 is 100, 5 is 101, and 6 is 110. Note the scheme of just adding one to the LSB. If your sum is greater than 1, you just move over a digit and carry the 1, like 9 in the decimal system. It is fun to practice adding binary numbers, then subtracting, and finally short and long division. Don't be surprised if your children, with their superior knowledge of new mathematics, make you seem to be an idiot.

Let's now list all binary equivalents of numbers through 15. All of the present slow-scan digital designs are limited to bits of 15 or less. Experimenters such as W6MXV have occasionally used binary equivalent numbers up through 32 but circuit complexity usually makes it impractical for the average experimenter.

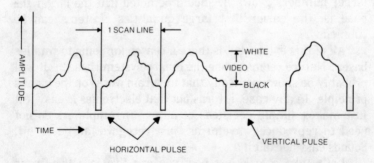

Fig. 6-1. A typical video signal. Note that each pattern is separated by a sync pulse, either horizontal or vertical. The portion of the pattern above the beginning of each sync pulse to the top is the video, ranging from black to white. There are either 120 or 128 horizontal pulses between each vertical sync pulse. Scan time for an SSTV picture is 8 seconds.

The list is as follows: You should make sure that you completely understand these numbers before going any further.

Decimal	Binary
0	0000
1	0001
2	0010
3	0011
4	0100
5	0101
6	0110
7	0111
8	1000
9	1001
10	1010
11	1011
12	1100
13	1101
14	1110
15	1111

Just remember that the circuitry will never be asked to do anything except to recognize an *on* or *off*. Relays or switches can do this, but it would be difficult to synchronize hundreds of people all throwing switching under a director or coach. Nevertheless, there is a director in digital circuitry, and its name is *clock*. This clock makes all the circuits switch together. There is much about this later.

We are so used to seeing analog-voltage waveforms displayed on an oscilloscope that you must condition yourself to not expect this kind of display in the digital world. Let's take the example in Fig. 6-1. This is a TV analog signal that could come from your TV camera. In this picture you can see the sync pulses as well as the monochrome portions of the picture.

Note that this particular pattern repeats. In TV the repeats represent scanning lines that are synchronized by negative-going pulses in the waveforms. In addition to the horizontal pulses shown, there are much larger pulses that occur at the start of each field or frame times called vertical sync pulses. In fast-scan TV this occurs every 1/60 of a second, but in SSTV it only happens once every 8 seconds.

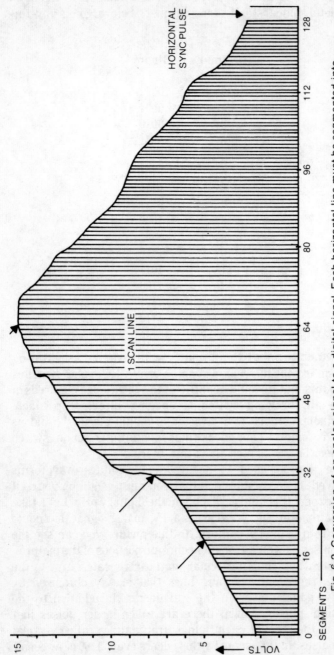

Fig. 6-2. Conversion of an analog signal to a digital signal. Each horizontal line must be segmented into 128 divisions. Arrows indicate the 5V, 8V, and 15V levels. From the black level, 0V, to the white level, 15V, there are 16 levels of gray. Gray levels can't exceed 16 because we have elected to use a 4-bit word system. Note that 0 is a level, thus 0-15V provide 16 separate levels. Any value between two voltage levels is considered to be the lower level (e.g. 5.6V is considered simply 5V).

How can you make this pattern into square waves of *ons* and *offs* in order to retain complicated TV waveform? You must first ask yourself how often you want to change this analog signal into a digital representation. In TV this usually means the following: What is the best resolution you can get with a given amount of memory (money)? As an example, let's say that it is desirable to do this conversion 128 times between horizontal sync pulses. For convenience, the number of items (segments) you use to divide each scan line should be divisible by 2—128 is 2^7. This can be represented as 128 divisions of each scan line in Fig. 6-2.

Your job now is to convert each one of these voltage segments into a digital representation. Take a typical segment of a scan line in Fig. 6-2—the oscilloscope shows this to be 5V. By this time you should instantly recall that 5 is 101 in the binary number system. The TV signal may go to 9V—this is 1001. Finally, the highest voltage found in the waveform is 15V—this is 1111. The MSB represents 8V; the LSB represents 1V. In this case we elect to consider any voltage between 1V and 2V to be counted as 1. The gray levels can not exceed 16 because we have chosen to limit ourselves to only 16 levels (0−15) and permitted only 128 binary conversions in each scan line. All of the analog voltages can thus be represented by binary numbers of four bits.

The reason for using 128 segments is that this number is compatible with the bandpass of SSTV video. Since each scan line takes 1/15 of a second to traverse the monitor screen and the segment frequency is 128 per scan line, then it follows that the bit frequency is equal to 1920 per line. This number (1920) falls approximately midband between 1500 Hz and 2300 Hz.

The best circuit to convert analog signals to binary numbers is the parallel A-D converter. The term parallel means that all bits (4-bit groups) associated with each segment are processed simultaneously. A bank of comparators is used to develop the signals to be encoded. As an example of the application of the comparators, consider the 1-bit A-D converter. Let's say that our video has a peak value of 15V. The A-D circuit is shown in Fig. 6-3. Our purpose in building this simple A-D converter is to develop 1-bit TV pictures; the picture will have two shades of gray (2^1). This is good enough for printing, handwriting, or machine copying.

If the video is below 8V, the comparator output goes to 0V; if the video is above 8V the comparator output is approx-

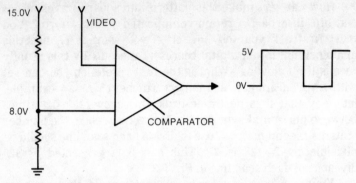

Fig. 6-3. A simple 1-bit comparator for A-D conversion. The output is 0V if the input video is below 8V; if the input video is 8V and above the output is 5V. The picture would have two shades of gray, or 2^1.

imately 5V—it is just that simple. Note that all amplitude variation of the input signal has been completely eliminated. Now this same idea can be extended to 2 bits, 3 bits, or 4 bits. It takes one less comparator than the number shades of grays desired. A 4-bit system has 2^4 shades of gray, so it takes 15 comparators. If dual comparators (711s) are used, then only eight units are required. The extra comparator is usually used for an over-voltage indicator.

In order to make sure that the idea is well developed in your mind, let us consider the 2-bit comparator shown in Fig. 6-4. Here each comparator is referenced to a different voltage. Outputs from this circuit are as follows: If the video level is less than 4V, A, B, and C output are 0V; if the video level is greater than 4V but less than 8V, A output is high with B and C outputs remaining low; if the video level is greater than 8V but less than 12V, A and B outputs are high with C output remaining low; if the video level is greater than 12V, A, B, and C outputs are high.

Now it is needed to develop a binary encoder. The entire purpose in going to a binary number system is to develop a simple representation of the original analog video signal. It is needed to encode the four shades of gray at the outputs of this comparator into just two bit groups. This is the purpose of the encoder.

In the 2-bit A-D comparator shown in Fig. 6-4, the smallest voltage that we can represent is 4V. This will be the level of analog voltage that will represent the LSB. Comparator 2 splits the input voltage at 8V which represents the MSB. There are four levels of output that can be used directly, but there is a much more efficient way of coding these four signals so that they can be represented by only two binary numbers. As an example, the four combinations of comparator outputs can be encoded by a 2-bit binary word encoder as follows:

Volts	MSB	LSB
0	0	0
4	0	1
8	1	0
12	1	1

Take a look at the requirement of the encoder, and see if we can design a simple 2-bit encoder. Remember that the LSB represents 4V. This could occur at any point on the input analog waveform. It is obvious that comparator 1 will have an output if the input voltage exceeds 4V. In binary logic this is represented by the symbol $A\bar{B}$. If the input voltage is between

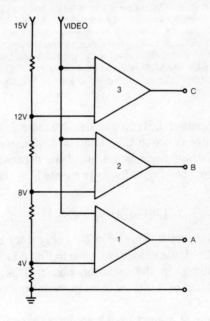

Fig. 6-4. A 2-bit comparator. An encoder is needed to make this circuit an A-D converter. Outputs from this circuit are as follows: if the input video is below 4V, all outputs are 0; if the input video is at least 4V but less than 8V, there is a 1 output from A, and B and C are 0; if the input video is at least 8V but less than 12V, there is a 1 output from both A and B and a 0 output from C; if the input video is above 12V, there are 1 outputs from A, B, and C.

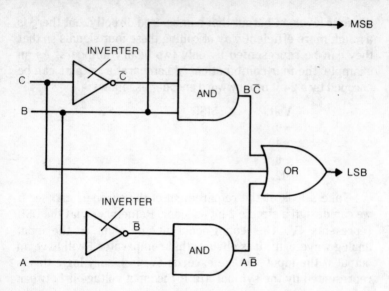

Fig. 6-5. A 2-bit encoder. Together with the 2-bit comparator in the preceding figure, this encoder output may be used to form 2-bit binary words. Combining this encoder with the comparator forms an A-D converter. Translated through these two circuits, each video voltage level becomes a 4-bit word. That is, if the video level is 0V, the MSB and LSB are both 0; if the video level is 4V, the MSB is 0 and the LSB is 1; if the video level is 8V, the MSB is 1 and the LSB is 0; if the video level is 12V, the MSB and LSB are both 1.

8V and 12V, another LSB can be designated $B\bar{C}$. And finally if the input voltage is above 12V, it can be designated just C. Note that the LSB designation is not all of the representation. The LSB can now be logically represented in the following equation:

$$\text{LSB} = A\bar{B} + B\bar{C} + C$$

The MSB is just B. This bit occurs only if the input voltage is above 8V. The logic design is shown in Fig. 6-5. The inverters give the $\bar{B}$ and $\bar{C}$ in the equation above. This idea can be extended to 3 bits or 4 bits as done in most amateur slow scan A-D converters.

We are now at a point in our analysis that we can begin to talk about memory requirements—SSTV has many reasons to store data. Remember the original 7290 vidicon? It stored an analog picture on a mosaic inside the vidicon; now we have digital bits to store. From the preceding analysis, we must store 128 × 4, or 512, bits for each line. The 128 *pixels* are the time division of the horizontal line and the 4 is the 4-bit group for each of the 128 pixels. This means that each line contains 512 bits to remember; therefore, a picture consisting of 128 lines will need 512 × 128, or 65,536 bits of storage. This memory capacity meets slow-scan standards nicely and can be used for fast scan if each of the 128 lines is displayed twice to give a picture consisting of 256 vertical lines displacement.

Now that you basically understand how to convert an analog signal to a digital one you may proceed ahead to see how useful this new technique is in the generation and display of TV pictures.

Up until now we have only said that we need a way to store all of the *on* and *off* digits, or more properly *1* and *0*. Computer hams should know of several ways to store bits. The only useful ones to TV applications are the methods that can work at relatively high frequency.

As an example, we decided that 128 horizontal time divisions, or pixels, would be used for each horizontal TV line. The number is not optimum but represents an economical trade off. In ordinary fast-scan TV, each line takes 1/15750 second or 63.4 μs to cross the screen. So it is easily calculated that $128/63.4 = 2$ million pixels per second. The same 128 pixels in SSTV must be scanned in 1/15 ms. This is seen to be 1920 pixels per second. These are the ratios of frequency difference between fast scan and slow scan. Any memory devices we use must be able to be clocked at frequencies approaching these extremes.

There are three devices on the market that meet these requirements: one is dynamic shift registers, another is RAM (random access memories), and the last is charge-coupled devices (CCD). The CCD *crystals* are expensive and have problems at low clock rates. The RAM memories have the distinct advantage of individual memory-cell addressing. These devices are just coming into general use now.

The memory devices that are available at reasonable cost, either surplus or new, are dynamic shift registers. These devices are connected in serial fashion to store one of the individual bits, that is, the MSB. A memory system for an entire 8-second SSTV field of 65,536 bits stored in 4 strings of 16 1024-bit shift registers is shown in Fig. 6-6. Since we have decided on a 4-bit digital word (16 shades of gray), all 4 bits result in 4 identical memories circulating in synchronism.

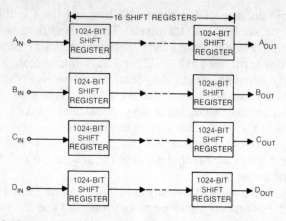

Fig. 6-6. Four identical shift-register circuits comprise a 4-bit serial memory containing 65,536 bits. This memory is divided into 4 serial chains containing 16 shift registers. Each chain is capable of storing one of the four bits on the digital word.

We are now ready to design a digital camera and scan converter. Our objective is to use an existing fast-scan CCTV camera and TV set as a monitor. Your old-time slow scanners will be happy to find out that you will not need to sit quietly for 8 seconds while your image is being televised or be required to look at a yellow P7 screen that loses its image quickly.

Let's review the properties that make all of this possible once again. A sample of video is broken into 4 *on-off* signals. Each is clocked into flip-flop circuits or shift registers that can store a bit of data. The use of an A-D converter, a memory, and a D-A converter makes all of this possible. The A-D converter changes the analog sample into a 4-bit code, the memory stores and changes the rate of the data passage, and finally the D-A converter turns it back into an analog signal ready for viewing on a monitor at a new scanning rate. This means fast-scan TV has been converted into an SSTV picture.

The digital camera can operate with or without large memories, but the slow- to fast-scan converter must have a large memory to hold that data which is coming in at a much slower rate than the display rate. Since the camera is operating at a very fast rate (60 fields per second), a single line out of 128 fast-scan fields can be stored temporarily and transmitted slowly. This type of simple digital camera still requires the operator to sit quietly for the full 8 seconds since one line is taken from each of the 128 fields. This is the principle of the line converter made popular by W0LMD and W6MXV.

A much better scheme is to store all 128 lines (approximately every other line on a conventional fast-scan field) in 1/60 second, then read on the stored lines at the slow-scan clocking rate. This requires the same size memory as the slow-to fast-scan converter.

The basic mechanism operating in any form of digital scan converter is the ability to store (clock-in) data at a given rate (2 MHz) and to clock out this same data at a much slower rate (2 kHz). This is similar to using a tape recorder operating at two speeds; one of these speeds is 1000 times slower than the other. One advantage in the tape machine is its ability to operate in the analog domain.

There is a similar device that may play an important part in analog scan converters. These are the CCD crystals; they are analog shift registers. These new circuits are revolutionary in their application to slow-scan problems, but they are very expensive.

Let us now assume that we have converted a normal 63.4 μs fast-scan TV line into 128 pixels with 4-bit words. Reviewing once again, this means 128 divisions along the horizontal line and 2^4 shades of grey. It is much better to have 256 pixels and 2^5 shades of grey but the cost of memory and circuit complexity limit the design. Clocked in 63.4 μs, 128 pixels results in a 2 MHz clocking pulse rate. These same 128 pixels with 4-bit words are now clocked out at a rate to be compatible with SSTV. This works out to be 128/66.6 ms or about 1920 Hz. This gives instantaneous fast-to-slow conversion on a line-to-line basis. This is truly a step toward digital space-age electronics.

What do you now do with 4-bit words of *on* and *off* pulses? You need a D-A converter to be able to see the resulting picture on a TV monitor. This circuit is the simplest of all. Let's assume that we have a video source with a maximum level of 15V. Converting the 15 to a binary word is expressed as 1111 by the A-D converter. The far left bit is the MSB and represents 8V; the next bit to the right represents 4V; the next bit to the right represents 2V; and the LSB represents 1V—expressed as $8 + 4 + 2 + 1 = 15$. Remember that any bit within each 4-bit word, regardless of its importance, is the same voltage height as the other 3 bits; that is, all 4 bits are

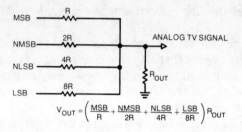

Fig. 6-7. A simple 2-bit D-A converter. The **N** in the terms NMSB and NLSB stands for **near**. These resistors are connected so that the current through each is proportional to its importance. The equation in the figure is nothing other than Ohm's law.

one voltage. If four resistors are connected so that the current through each of them is in proportion to their importance in the binary word, we have designed a D-A converter as in Fig. 6-7.

Isn't this simple? You are now back in the analog world after a short trip through digital electronics. Two things to remember are (1) if you have to convert fast to slow scan, connect the output of the D-A converter to the SSTV monitor; (2) if you have to convert slow to fast scan, connect the output of the D-A converter to the input of the video amplifier.

In the chapters to follow you will see how you can do all of this in color, so you can be looking at color slow scan in the living room; the screen will be a color representation and the scene will change once every 8 seconds. The block diagram of a simple camera and slow- to fast-scan converter is shown in Fig. 6-8.

W0LMD DIGITAL SCAN CONVERTER

The digital slow- to fast-scan conversion process results in a completely new generation of slow-scan art. This process may be defined as capturing and viewing the video in real time, while actual transmission takes place over 8 seconds.

Several systems are possible to achieve the objectives of second-generation SSTV. Storage tubes for receiving and Lithicon storage tubes for both transmission and reception scan conversion are potential alternatives, but their high cost and lack of total system flexibility tend to diminish their value.

The alternative digital system requires a change to a vastly different point of reference. This system requires very sophisticated logic elements to enable the complex component

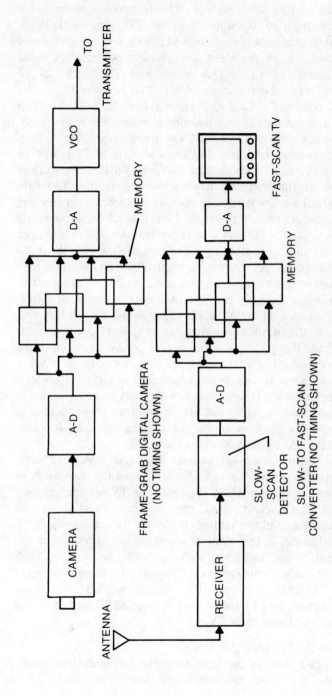

Fig. 6-8. Simplified block diagram of slow- to fast-scan converter for displaying SSTV signals on a fast-scan TV. The upper portion of the diagram is the camera setup required for frame grabbing.

119

elements to function together. The technology embraces a wide spectrum of IC ships, such as TTL and MOS. Logic elements include not only gates and clocks, but also such items as dynamic shift registers, two-phase clock drivers, code conversions, and A-D or D-A conversions. The impact on the ham wishing to become involved in this art is obvious.

The possibility of a digital video system for SSTV has been known and discussed for a number of years. The major problem has been the memory cost and complexity. Slow- to fast-scan conversion requires that the complete SSTV frame be memory stored at a given time. Several considerations affect the size of the memory needed for one frame. The first consideration is the definable number of elements per line and the number of lines. At this time, I feel that 128 definable pixels per line times the 120 SSTV lines per frame would be required for minimum acceptability. This multiplies out to 16,384 definable points per frame, at least. Since SSTV is not only black and white, but also gray values, a digital system must be built to define the gray component of each of the 16,384 points. Sixteen shades of gray, each represented by 4-bit words, illustrates the minimum acceptable gray level quality for the total 65,536 bits per frame. Since cost per bit has been in the neighborhood of a few cents per list, the cost of even a minimum system has been a prohibitive factor. Recently, however, memory IC costs have drastically fallen to a point of being less than a penny per bit. In addition, these memories are beginning to show up on the surplus markets. IC chips needed for a 65,536 memory cost about $30.

The scan-conversion project may be developed in a number of sequential steps. *Phase I* gives the dimension of "load on the fly" in which the fast-scan TV set continuously displays the picture, and the memory updated as the recirculating picture timing matches the incoming SSTV picture timing. In this system, a picture would remain on the fast-scan TV set, and picture change would be shown as each updating SSTV picture arrives. *Phase II* adds the dimension of frame grabbing *Phase III* the dimension of image enhancement, and *Phase IV* that of color SSTV viewed in real time on your home color TV set.

Load-on-the-Fly Scan Conversion

Since a slow- to fast-scan converter has many interrelated circuits, the usual block-by-block description would fail to bring out these interrelationships. Therefore a level, presentation has been utilized, starting out with an overall perspective of the operations, then a specific description of the logical elements, and finally a detailed presentation of component functions.

The primary function of this slow- to fast-scan converter (Fig. 6-9) is to convert the incoming SSTV signal into a form viewable on a standard monochrome TV set. As one choice of several alternatives, this unit uses a digital process, composed of digitized *bytes* of video data and digital-memory IC chips, as its primary conversion medium.

Several alternatives within this digital process are open to you; one alternative is presented here. This alternative converts the incoming SSTV video data from each SSTV scan line into 128 digital samples and loads each *odd* line into a 128-byte odd-line memory buffer and each *even* line into a 128-byte even-line memory buffer. Each buffer is capable of holding one complete SSTV digitized line.

A main memory—65 thousand digital bits worth of shift registers arranged in 4 banks of 16 thousand bits—is constantly fraction recirculating every 1/60 of a second. By keeping track of the precise address of this recirculating memory, as well as providing an ability to write to and read from this memory, the line buffers can update the appropriate section of main memory as it swings by. The readout function of main memory is made constantly available to video output and D-A sections for display on the TV set.

This ability to update this recirculating main memory from the incoming SSTV means that a picture will be continuously displayed on the TV set. The only time any apparent video change occurs is during a newly received video frame. In fact, after a good picture is received, the SSTV may be switched off, and the picture will remain until the scan converter is turned off, or until the converter receives some more SSTV signals which would overlay the recirculating video data. The memory can be completely updated in one 8-second SSTV frame. The operation appears identical to P7 reception, except that the previously written lines are just as bright as the line begin written, and there is no remnant of the old picture due to phosphor video integration. Once a line is updated the old video information is long gone. The brightness, definition, and detail resolution are vastly superior to any P7 monitor; even outdoor operation is entirely possible.

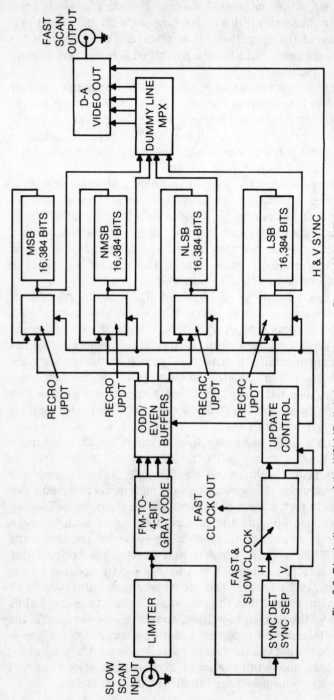

Fig. 6-9. Block diagram of the W0LMD digital scan converter, Phase I, Mark I. This converter was designed by Dr. Robert Suding and uses 16 shades of gray with 128-pixel horizontal lines. The memory is constantly recirculating and updated.

The complete operation then revolves around the ability to precisely update this recirculating memory. Because of the obvious lack of any direct speed relationship between the incoming SSTV and the output fast-scan TV, two separate clocks are utilized—a slow one to control video data movement *into* the buffer registers; and a fast clock to control video data movement *out of* the buffers into main memory, recirculate main memory, and output video to the fast-scan TV set. The buffers must then be the speed linkage—the ability to run slow, fast, or not at all—without ever accidentally losing any video data. Controlling the buffer operation will then allow a constant picture instead of slowing the main memory and losing video on the TV set.

The buffer size can have many possibilities; it could be as long as main memory, accomplishing update in one crack; it could be as short as ¼ line, updating a line at a time; or it could be anywhere between these extremes. A *complete* frame buffer is a waste of memory; it takes 8 seconds before anything happened on the screen, reducing your confidence while viewing the screen. A ¼-line update is feasible, but it requires quite sophisticated logic to insure smooth blending of the video line segments. The best compromise is a complete line buffer.

Main memory IC chips can be a number of different IC types, but they must work very dependably at 3 Mhz under sequential operation without ever losing a bit. This imposes a very stringent requirement on the IC chips and eliminates a large number of possibilities. The obvious choice is an LSI MOS dynamic shift register with around 1024 bits per chip. These units are available from a number of manufacturers in various forms. Most of these devices use an external 2-phase clock, one phase of which moves data *in*, and the other moving the data *out*. A 1024-bit shift register would then take 1024 phase-1 pulses and 1024 phase-2 pulses, alternating each to accomplish data movement through this one IC. Since this particular design uses close to 16-thousand-bits worth of IC chips (16 groups of 1024 bits) in each (4) parallel bit planes; this means 32 thousand alternating clock pulses. IC shift registers may be multiplexed for HF operation, and some IC devices are designed with on-chip multiplexers, resulting in lower clock-pulse counts for the same data rate movement.

The 2525 shift register is the essential memory element of this particular design. This is nonmultiplexed chip capable of a

5 Mhz data and clock rate. A common alternative to the 2525 is the 1404 that has a data frequency limit of 10 Mhz due to on-chip multiplexing. This chip would work very well in this circuit, even better than the 2525, except that the clock rate necessary would have to be ½ the speed, which could be accomplished by tapping the drive to the 2-phase clock generator one position down the frequency-divider chain. The major criterion for chip selection is to find the cheapest that will do the job. In this particular design, 64 1024-bit IC chips, plus 20–30 spares, are needed; so at $9 each this gets expensive! Nevertheless, when purchasing 400 2525 chips for 25¢ each (surplus "as is"—about 50% checked out okay), the decision for the 2525 is obvious. Unfortunately sources dry up quickly, and so you must keep an open eye and hang loose. The other supporting logic will run around $150, so the cost of the memory IC chips will be a key factor in the overall system cost.

The receiving TV set can be any monochrome TV set, but since a large screen requires distant viewing only, some moderation is in order. Relying on small P7 screens is not valid in this system. Viewing from 50 feet away is possible using a 19-inch Zenith, and a 9-inch 311 Philco (my preference) provides good viewing from 15 feet.

Several quality enhancements are possible and recommended: The aspect ratio difference is a problem that can be solved by gating the fast clocking of main memory such that video output is sent only to the receiving monochrome TV set during a middle portion of each horizontal line, sufficient in length to result in an essentially square picture. Aspect blanking is provided so that the video on the screen is enclosed by a black border on either side of the picture. Since the incoming SSTV signal is 120 (or 128) lines per frame, an obvious and distracting dark line exists between each line when viewing on a large and bright TV screen. To avoid this problem, a dummy line doubler is provided so that each SSTV line is outputed to the fast-scan TV set twice, greatly enhancing the video image.

FM-to-Gray-Code Converter

The input MC1741SC (Fig. 6-10) acts as a fast limiter/squarer, producing about a 20V P-P square wave. The input to the MC1741SC is clamped by a pair of diodes that limit the input voltage to ±0.6V to protect the IC.

The output of the limiter then goes to two sections of a 7400 which functions as a driver/squarer for the differentiator. This circuit has been adjusted to insure 1/2-cycle balance.

The differentiator drives a 74122 retriggerable one-shot delay. Each cycle will fire the 74122 at an adjustable period of time, from 400 μs to 500 μs, ideally about 440 μs. This time duration corresponds to the cycle duration of a 2300 Hz *white* signal on SSTV. At the end of the delay, the 7413 clock is allowed to run; the resultant positive *rise time* increments the counter following it.

The clock is also variable in speed from 54 kHz to 82 kHz, ideally about 70 kHz, which will produce 16 counts in the 1/2-cycle difference between 2300 Hz and 1500 Hz (116 μs). The 7413 clock drives the 74193 quad-binary-synchronous counter that is triggered on the positive-going pulse edge. The further down the incoming signal is from 2300 Hz, the longer the clock will be running, and the greater the count value is in the 74193—ranging from 0 with a 2300 Hz signal to 15 for a 1500 Hz or lower signal—1 count more for each 50 Hz below 2300 Hz.

At the end of the cycle, a third section of the 7400 will fire a strobing pulse that loads the count value into the 7475 memory. Slightly later, the fourth section of the 7400 sends out a pulse to the 74193 to reset it to 0 in order to prepare it for the frequency check on the next cycle.

The second half of the 7413 checks the count at the output of the 74193. When a binary value of 15 is at the output (all 1s), the 7413 degates the 7413 clock, thus preventing a count-overrun condition on some overblack signals, or slightly mistuned signals. Sync time is also blanked black in this way.

The 7475 outputs go to three sections of a 7486 that convert the binary output to the gray code, which permits data transfer into memory without strobing, and preventing apparent false bits from showing up after data reconversion.

LED devices (MV5020) are used to show when no bits are strobed into memory (2300 Hz.), or when all 4 bits are in the memory (1500 Hz or less). To prevent sync from also lighting the *black* LED, the sync-detection circuits degate the black indicator's logic. Since there is a slight delay in the rise time of the SSTV horizontal sync pulse, dim blinking of the black LED results, giving an indication of sync detection, though still enabling the operator to effectively adjust the black and white compression controls.

Slow-Scan Sync Detection

The slow-scan sync sections (Fig. 6-10) are designed to utilize a highly modified version of the W0LMD sync circuitry. The basic idea is to use a threshold-detection method instead of the usual tuned-circuit approach. The advantage of the threshold system is that the actual sync frequency is unimportant; whatever signal is the lowest frequency is defined as sync. This allowed the monitor to be as much as 200 Hz off frequency—high or low—without losing sync; this is obviously of great value in net operations for hands-off reception of pictures if the audio is understandable. The very wide passband has been found to be a disadvantage, however, since any interference below the sync frequency can be interpreted as sync also.

To maintain the advantages of hands-off tuning and eliminate interference problems, a specially tuned circuit and threshold detector is utilized in this circuit. The tuned circuit is designed to present a controlled detection slope to the threshold detector. The slope is established so that interfering signals outside a narrow range will have no effect on the threshold system. This is accomplished by setting the tuned section to peak at approximately 1100 Hz and have essentially no output at 1600 Hz; this falls within the previous sync tolerance of ±200 Hz.

The output of the limiter is fed to an active bandpass filter tuned to about 1100 Hz, with a gain of 0.65 and a bandwidth of 200 Hz. The high-edge slope then gives a diminishing output as the frequency rises above 1100. Thus, sync (unless tuned well below 1000 Hz) will always have a greater amplitude than video frequencies.

A full-wave IC detector converts the AF from the BPAF to a full-wave positive-pulsating voltage for the AST. The AST (auto sync threshold) permits the pulsating voltage to charge two capacitors at the input. The 0.039 μF capacitor merely reduces the pulsating nature of the incoming signal, while the 10 μF capacitor actually charges up to the most positive value of the incoming signal, which is sync, and holds this voltage for reference. A variable portion of this reference voltage is applied to the noninverting leg of the operational amplifier while the full voltage incoming goes to the inverting input. Now, any time sync comes in, the output of the AST switches to minus, while video does not break the threshold. If the incoming SSTV signal is off frequency, the reference voltage changes of course, but the sync-video relationships remain the same. Sync frequencies of 1200 Hz ±200 Hz copies quite well.

Separate low-pass filters are used to detect the presence of vertical and horizontal sync. The threshold is variable to eliminate false pulses but still pick up sync under fading conditions. Their outputs are converted to TTL levels by a pair of Schmidt triggers and drive update process control, as well as controlling the slow-clocking circuit.

Clocking

The slow clock (Fig. 6-11) is designed so that it is synchronous with the incoming SSTV signal. The slow clock is gated *off* with no incoming SSTV signal. When a horizontal sync pulse is detected, however, the slow-clock counters are reset to 0, and a gating-off level is removed from the slow clock. The 7413 clock then free runs at approximately 2130 Hz, and its output is coupled both into the buffer update logic on board 3 (Fig. 6-12), and into a pair of 7493 counters. After 128 counts, pin 11 of the second 7493 (Fig. 6-11) goes positive, and after inversion the 7413 is gated off until the next incoming horizontal sync pulse resets the counter, and the process begins again. The speed of slow clock may be adjusted for aspect ratio and may be increased to make 50 Hz stations fill out the picture area.

Fast-scan-line timing is controlled by two 74193 chips, the first acting as *divide 12*, and the second as a *divide 16*, for a total of *192 divisions*. The first 128 counts are used to control SSTV picture-painting time on the TV set, while the other 64 counts allow blank portions at either side of the resultant picture for a more nearly square picture.

Three time decoders are involved in the *divide 192* line-timing section. The first section is a 7400 and an inverter, resetting the counter to 0 after 192 counts. The other two time decoders establish the fast-scan horizontal sync pulse in about the middle of the aspect-ratio blanking-64-count period, approximately centering the SSTV picture on the TV set. The fast-horizontal-sync-time decoding gives approximately 7 μs pulse. The 7420 decodes *time* 148 to set an R-S flip latch *on*, and the 7400 section detects *time 168* to turn the flip latch *off*.

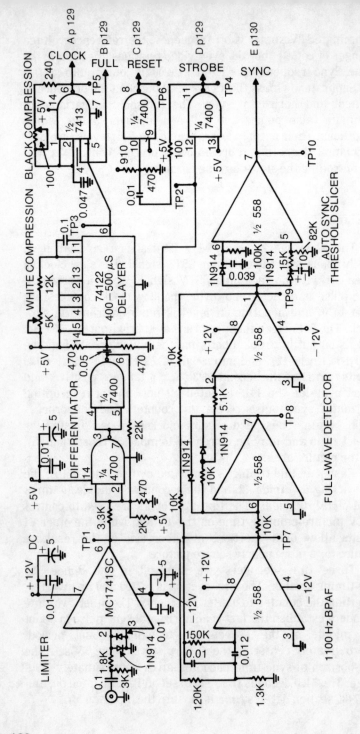

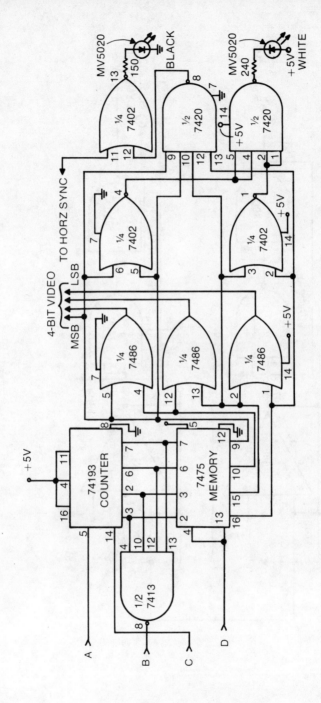

Fig. 6-10. Schematic diagram of PC board 1 in the W0LMD slow- to fast-scan converter. This board contains the FM-to-gray-code converter and the sync circuits.

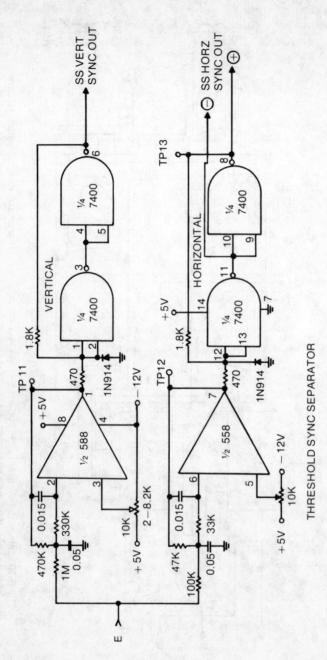

Fig. 6-10 con't

Fast-scan frame timing is controlled by two more 74193 chips each functioning as a *divide 16* for a total of *256 divisions* the number of lines which will result from this scan converter to the TV set. The 7420 decodes *time 240—256*, which results in a 1 ms fast-scan vertical sync pulse. Since half of the lines are dummy lines, gating from the first frame-counting 74193 identifies the line as coming from the main memory or from the dummy buffers.

The two-phase clocking runs the 64 1024-bit shift registers by delivering 128 phase-1 pulses alternating with 128 phase-2 pulses, during the period of line time when main memory is delivering an SSTV picture to the TV set. The two phase-clocking circuit derives alternating pulses by ANDing delayed and inverted pulses coming from the crystal clock. A 74L04 is a slower member of the TTL series that allows selecting the desired delay and resultant two-phase pulse widths with a minimum of parts.

The actual crystal frequency is determined by multiplying the frame rate (60) by the number of lines (256) and the number of picture elements (PE) per line (192) for the resultant 2949.12 kHz. Alternately, the standard fast-scan horizontal frequency (15750 Hz) may be multiplied by the PE/line (192) for 3024 kHz. Either will work quite well, but the first calculation requires trimming the TV set's *horizontal hold* slightly, while the second requires trimming the *vertical hold* control. The lower frequency runs the memories slower and hence, slightly more conservatively, while the higher frequency gives a nearly square picture. Actually, a crystal of 3 MHz ± 200 kHz will work well.

Buffers and Updating Control

The brains of the scan conversion process is located on board 3 (Fig. 6-12). This board contains the two line buffers for loading the main memory from SSTV and the supporting switching logic. The incoming SSTV vertical sync pulse resets the 7493 chips to 0, then subsequent horizontal sync will increment the count value residing in the 7493 chips such that their binary value is the SSTV line number currently being received.

The fall of the SSTV horizontal sync pulse starts the odd or even buffer (depending on whether the incoming SSTV line is odd or even), loading at the 2130 Hz rate, resulting in 128 samples of the incoming video being loaded into the

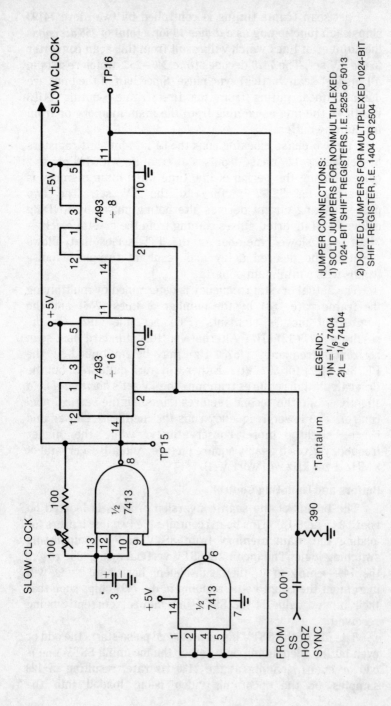

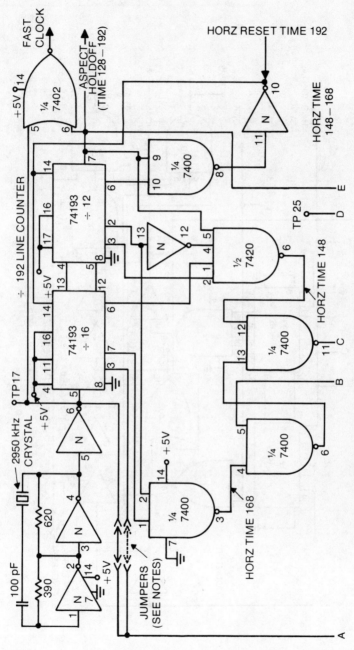

Fig. 6-11. Schematic diagram of PC board 2 in the W0LMD slow- to fast-scan converter. This board houses the slow and fast clocks.

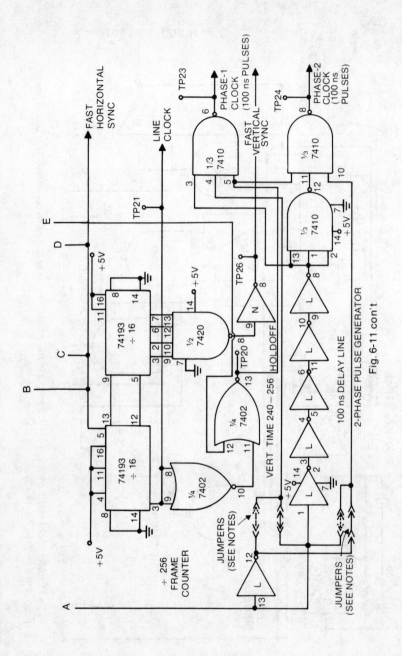

Fig. 6-11 con't

appropriate line buffer. At the fall of each SSTV horizontal sync pulse, the buffer not assigned to the incoming SSTV line now outputs its previously loaded line into main memory when the appropriate section of main memory is cycling by. The detection of the correct section to be updated requires many special temporary memory registers, which necessitates precise timings.

The pair of 74193 counters are preset to the value residing in the 7493 line counters by a 500 ns preset pulse generated at the fall of each SSTV horizontal sync. This pulse also brings up a line called *update request*, which when ANDed with the rise of a fast-scan vertical sync pulse (indicating the beginning of a fast-scan frame), brings up a latch called *update enable*. With update enable *on*, the fall of each fast scan line will decrement the 74193 counters until a *borrow* occurs on pin 13. Obviously, the number of decrements before *borrow* occurs is directly proportional to the value to which the counters have been preset.

Borrow is called update after inversion and lasts the required 63 μs required to accomplish updating of main memory. During this interval the main memory bank is switched from recirculate to load from the appropriate line buffer clocked by the fast-clock pulses. When the line buffer is emptied, further unload attempts must be blocked until the next valid update request, so the update request and update enable are dropped for the duration of the SSTV line.

The actual placement of the update pulse relative to the SSTV horizontal sync pulse will be jumping all over, but this is only normal due to the need to perfectly synchronize the random timing of the incoming SSTV with the recirculating main memory. By looking at the update pulse while syncing on the update enable pulse, the smoothly syncronized sliding pulse updating process becomes evident.

Power Supply

The power supply requirements greatly depend on the type of memory IC chips utilized. The 2525 uses +5V and −5V supplies. Each IC will consume about 25 mA, approximately a 1.75A requirement on the supply from this source alone. The power supply (Fig. 6-13) must also furnish +12V and −12V.

A 24V center-tapped 2A control transformer is used in a dual center-tapped bridge configuration. A more desirable

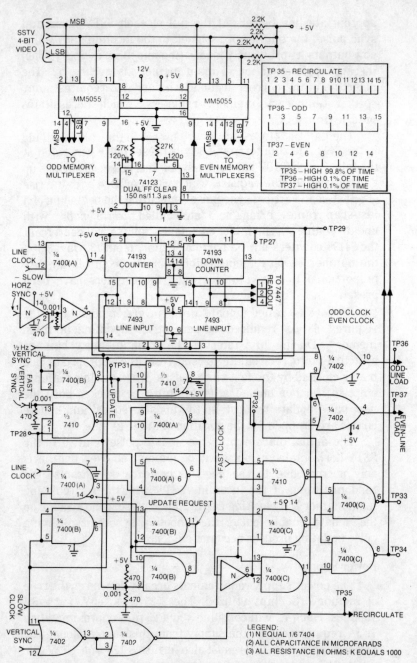

Fig. 6-12. Schematic diagram of PC board 3 in the W0LMD slow- to fast-scan converter. This board contains the load-on-the-fly slow/fast buffer control.

transformer arrangement would be a transformer for +12V and −12V supplies, and a 15V center-tapped transformer for the +5V and −5V supplies, which would result in considerably less heat dissipation in the regulators. The rectifiers supplying the +5V and −5V legs are 5A units. The +12V and −12V rectifiers do not have the stringent current requirements, so standard 0.75A silicon rectifiers are used. Overload protection is provided on all supplies, as is electronic regulation. Each supply voltage is set by a 3-terminal regulator. To boost the current capabilities on the high-current supplies, external series regulators are used in current-sharing circuits which retain the overcurrent and temperature-protection features of the 3-terminal regulator while maintaining simplicity.

Main Memory

Boards 4, 5, 6, and 7 (Fig. 6-14) are the main memory boards, with each board being dedicated to a single bit plane. Each board has 16 2525 shift registers mounted on Molex pins. Also on each board is a MH0026CG (recently relabeled an LH0026CG), a 7454, and a 7400.

The 16 shift registers on each board are wired in series for a bit total of 16,384. The inputs from the buffers are multiplexed with the output of the shift-register string by the 7454 quad mutliplexer. A single section of the 7400 controls the video from the memory board as well as drives the shift-register string.

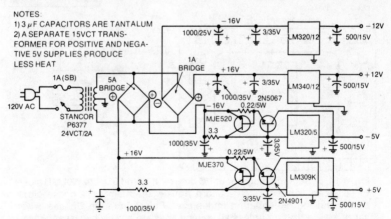

Fig. 6-13. Schematic diagram of the suggested power supply for the W0LMD slow- to fast-scan converter. A separate 15V center-tapped transformer for the +5V and −5V supplies will produce less heat dissipation.

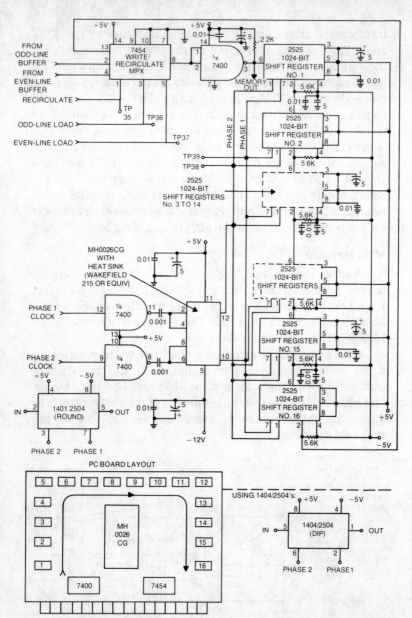

Fig. 6-14. Schematic diagram of PC boards 4, 5, and 6 in the W0LMD slow-to fast-scan converter. This board houses the memory, 2-phase clock driver, and the recirculating multiplexer. Pulses from the 2-phase driver cause severe ringing unless all voltage lines are heavily bypassed. A two-sided board is suggested, with the top as the ground at the bottom carrying the voltage supply lines. Place the MH0026CG in the center of the 16 shift registers.

The 7454 multiplexer selects the input source when that source has its selected line high. Since the majority of the memory time is spent recirculating previously loaded video data (99.8% of the time), the recirculate control line is high except for 63 μs pulses occuring at update time when receiving SSTV signals. At update time, the buffer whose turn it is to output will have a 63 μs positive pulse on its select line, resulting in the noncirculation of the previous data associated with this line, and the insertion into the main memory of 128 samples associated with the updating line.

The 2-phase-clock-driver system receives the pulses from the clock board (2) and develops fast-rise and fast-fall clock pulses capable of driving a heavily capacitive load. The clock lines to the shift registers should obviously be short and attached at multiple points on the shift-register-clock busses.

It is imperative that the recommendation for a large ground plane for bypassing be followed, or severe ringing will guarantee that the memory shift registers will not run properly.

D-A Converter

Board 8 (Fig. 6-15) in the present digital scan converter receives the video from the circulating main memory and produces an analog video output capable of being interfaced to a standard TV set. As with the previous design decisions, a number of alternatives would work.

The output from main memory is sent to some dual 128-bit shift registers. Each time main memory runs, it places 128 samples of the video line into these dummy line buffers and directly feeds the line to the D-A multiplier sections following. After the main memory has loaded the dummy line buffers, its video data movement will halt for one fast-scan line, but the dummy line buffers will continue to move, now outputting the video line just loaded. The 74157 multiplexer is switched over to the dummy line buffers to receive video data.

The output of the 74157 feeds a 7486 exclusive OR which converts the gray code to a binary code needed for proper D-A conversion. The scan converter was tried without the gray code encoder and decoder; however, a very significant improvement in picture quality was noted when using the gray code. The inclusion of gray code at this time also greatly improves the interfacing capability of this scan converter to the frame-grabbing feature being designed.

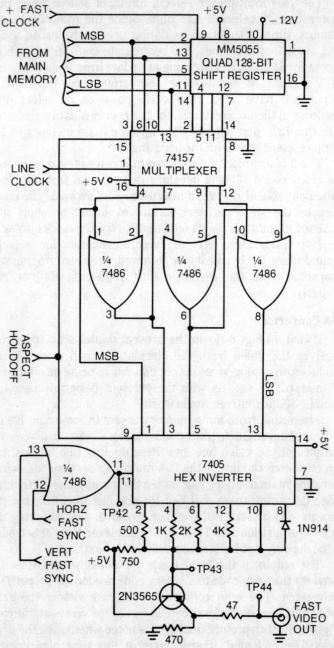

Fig. 6-15. Schematic diagram of PC board 8 in the W0LMD slow- to fast-scan converter. This board contains the D-A circuitry and the video-out stage.

Shift-Register Tester

This board (Fig. 6-16) was designed to test MOS shift registers, particularly surplus ones, prior to using them in the slow- to fast-scan converter. All of the voltages and pulses needed to check out the shift registers are provided. Power-supply design is left to the individual builder. Short-circuit protection is suggested for obvious reasons.

The unit allows data clocking at both high and low speeds. Often shift registers will work fine at one end of the frequency specifications but poorly or not at all at the other end. The LF speed may be adjusted by changing the value of the 1.5 μF capacitor. The HF speed is set by the 620 pF capacitor. The clock timing shown is usually sufficient to catch any defective unit.

Data is artificially generated by gating the clock pulses such that a seemingly erratic data stream is applied to the shift registers under test. This simulates the random nature of the data normally handled by the shift registers. A 2-phase clock signal is generated for use by the 1024-bit registers. Other shift registers requiring this type of 2-phase clocking may be tested by providing the proper voltages and pin connections.

Be aware that the MH0026CG will become warm when left in the high-speed position with 2-phase clocking enabled. Leave the speed switch in the 1.4 kHz position, and turn off the power when charging an IC.

The 7486 output sections enable you to readily detect missing or shortened pulses, particularly at high-speed clocking; the output of a known good shift register is exclusive ORed with the shift register in question. If both shift registers are good, the 7486 cancels the pulses, resulting in a straight line on the oscilloscope. Missing or shortened pulses show up as pulses on the oscilloscope. By using a dual-trace oscilloscope, both canceled and uncanceled patterns may be viewed.

Converter PC Boards

If you are interested in constructing this converter, PC boards are available from WA9UHV of Kokomo, Indiana for this unit. The WA9UHV kit consists of an SSTV input board, fast- and slow-clock board, update control board, four memory boards, an output board, and the special board for testing surplus memories. The current price of the kit is $65.

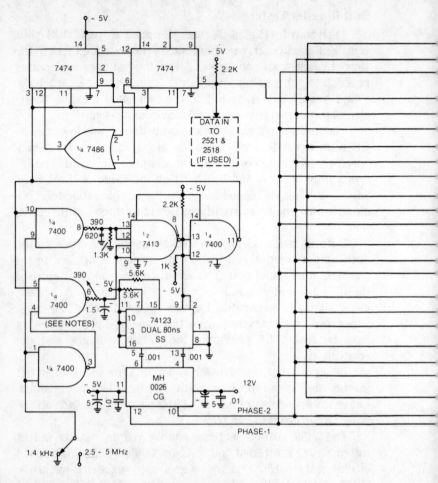

NOTES:
1) Change the value of the 620 pF capacitor to adjust high frequency.
2) Change the value of the 1.5 μF capacitor to adjust low frequency.

Test Point Waveforms

Waveforms shown in Figs. 6-17 through 6-23 are the correct oscilloscope patterns for the W0LMD slow- to fast-scan converter. They will be of great help in construction of this unit. Figure 6-17 contains the proper waveforms for the test points in the FM-to-gray-code section (TP1-6) on board 1. Figure 6-18 contains the waveforms of the sync detection and separation sections (TP7−14) on board 1). Figure 6-19 contains the clocking waveforms at TP15−17 on board 2. Figure 6-20 contains the waveforms for clocking and sync

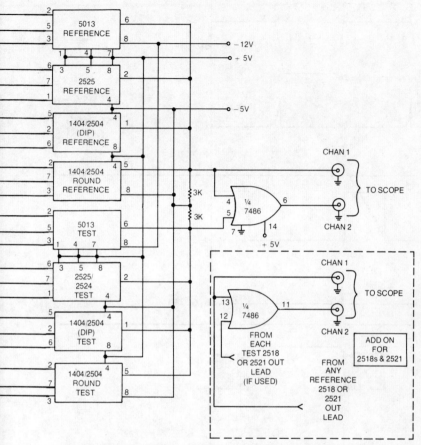

Fig. 6-16. Schematic diagram of the MOS shift-register tester. Test only one set of memory types at a time. This test board is used to check surplus shift registers before installation.

pulses at TP20, TP23, TP24, TP25, and TP26 on board 2. Figure 6-21 contains the waveforms at TP27, TP28, TP29, TP31, TP32, TP33, and TP34 on board 3. Figure 6-22 is the waveforms at TP35−41 on memory boards 4, 5, and 6. Figure 6-23 contains the waveforms for TP42−44 on D-A board 8.

MXV-200 SCANVERTER

Most SSTV cameras now in use fall into two general types: one type is designed and built specifically for SSTV and the second type is a fast-scan TV camera that has been modified to

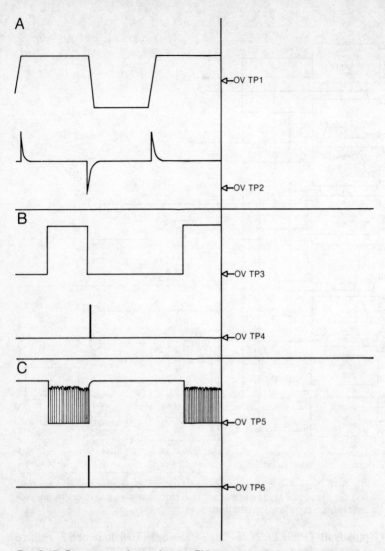

Fig. 6-17. Proper waveforms for the FM-to-gray-code section on board 1. (A) A dual-trace waveform at TP1 and TP2. TP1 is connected to channel 1 input with the following control settings: 10V/division, positive trigger, positive slope, 100μs/division, trigger on channel 1. TP2 is connected to channel 2 input with 2V/division. (B) A dual-trace waveform at TP3 and TP4. TP3 is connected to channel 1 with the following control settings: 2V/division, positive trigger, positive slope, 100 μs/division, external trigger on TP1. TP4 is connected to channel 2 with 2V/division. (C) A dual-trace waveform at TP5 and TP6. TP5 is connected to channel 1 with the following control settings: 2V/division, positive trigger, positive slope, 100 μs/division, external trigger on TP1. TP6 is connected to channel 2 with 2V/division.

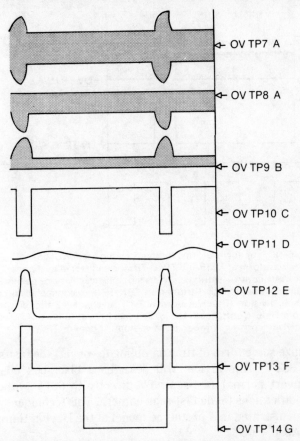

Fig. 6-18. Waveforms of the sync detection and separation sections on board 1. (A) A dual-trace waveform at TP7 and TP8. TP7 is connected to channel 1 with the following control settings: 5V/division, positive trigger, positive slope, 10 ms/division, trigger on channel 1. TP8 is connected to channel 2 with 5V/division. (B) A single-trace waveform at TP9 with the following control settings: 5V/division, 10 ms/division, positive trigger, positive slope, external trigger on TP7. (C) A single-trace waveform at TP10 with the following control settings: 10V/division, 10 ms/division, positive trigger, positive slope, external trigger on TP7. (D) A single-trace waveform at TP11 with the following control settings: 10V/division, 10 ms/division, positive trigger, positive slope, external trigger on TP7. (E) A single-trace waveform at TP12 with the following control settings: 5V/division, 10 ms/division, positive trigger, positive slope, external trigger on TP7. (F) A single-trace waveform at TP13 with the following control settings: 2V/division, 10 ms/division, positive trigger, positive slope, external trigger on TP7. (G) A single-trace waveform at TP14 with the following control settings: 2V/division, 20 ms/division, positive trigger, positive slope, external trigger on TP7 (no SSTV input). The single-trace waveforms of TP9 −14 are usually applied to channel 2 of the dual-traced oscilloscope while keeping the TP7 waveform on channel 1—note that all of the pulses are aligned vertically in the figure.

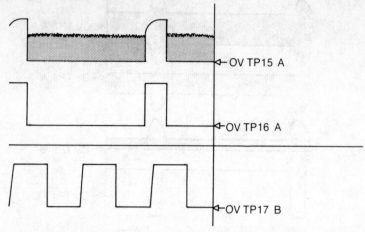

Fig. 6-19. Proper waveforms for the clocking section of board 2. (A) A dual-trace waveform at TP15 and TP16. TP15 is connected to channel 1 with the following control settings: 2V/division, 10 ms/division, positive trigger, positive slope, external trigger on TP7. TP16 is connected to channel 2 with 2V/division. (B) A waveform at TP17 on channel 2 after disconnecting it from TP16. Control settings for channel 2 are as follows: 2V/division, 100 ns/division, positive trigger, positive slope, trigger on TP7.

utilize some form of time sampling to convert the fast-scan TV to SSTV. This article will describe a black-box adapter to convert normal fast-scan TV directly to SSTV without any modifications to the fast-scan camera. This technique was first demonstrated in a prototype model at the Dayton Hamvention in April, 1973.

The technique used in the MXV-200 (Fig. 6-24) is to convert the fast-scan TV signal into digital coded information, store the TV picture in a memory at the fast rate, and then recall the TV information from the memory at a slower rate that corresponds to the SSTV picture rate.

For the purpose of description, the 525-line 60 Hz standards will be used; but the same technique may be used with 625-line 50 Hz standards used in other countries.

The 525-line standard consists of two fields of 262 half-lines, requiring 1/60 second to scan one field. The horizontal line rate is approximately 15750 Hz., with one line equal to 63.5 μs in time. The SSTV standards now in use are a field of 120 or 128 lines and a line rate of 15 Hz with a time of 66.67 ms. Thus, the fast-scan line rate is more than 1000 times faster than the SSTV line rate. The memory in the MXV-200 need only store one horizontal line at a time since the next

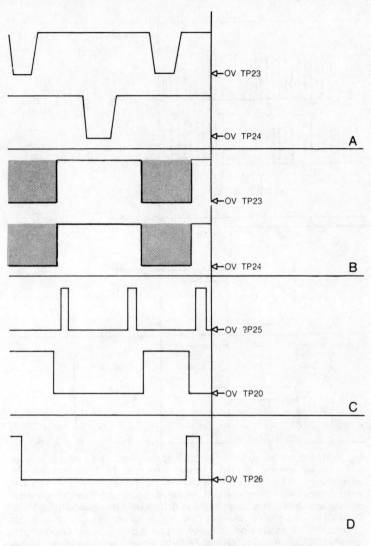

Fig. 6-20. Proper waveforms of the clocking and sync pulses on board 2. (A) A dual-trace waveform at TP23 and TP24. TP23 is connected to channel 1 with the following control settings: 2V/division, alternate sweep, negative trigger, positive slope, 50 ns/division, trigger on channel 1. TP24 is connected to channel 2 with 2V/division. (B) Same as above but with 20 µs/division. (C) A dual-trace waveform at TP25 and TP20. TP25 is connected to channel 1 with the following control settings: 2V/division, alternate sweep, positive trigger, positive slope, 10 µs/division, trigger on channel 2. TP20 is connected to channel 2 with 2V/division. (D) A waveform at TP26 on channel 2 after disconnecting it from TP20 above. Control settings for channel 2 in this case are as follows: 2V/division, 2 ms/division, positive trigger, positive slope, trigger on channel 1.

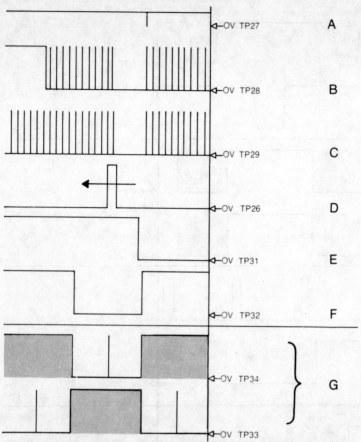

Fig. 6-21. Proper waveforms for the update control on board 3 with an SSTV signal applied. (A) The channel-1 trace at TP27 for **B** through **F** waveforms. TP27 is connected to channel 1 with the following control settings: 2V/division, 10 ms/division, negative trigger, positive slope, trigger on channel 1. (B) The lower portion of a dual-trace waveform at TP28. TP28 is connected to channel 2 with the following control settings: 2V/division, 10 ms/division, negative trigger, positive slope, external trigger on TP27. (C) A lower trace at TP29 of a dual-trace waveform. TP29 is connected to channel 2 with the following control settings: 2V/division, 10 ms/division, negative trigger, positive slope, external trigger on TP27. (D) A channel-2 waveform at TP26 with the following control settings: 2V/division, 2 ms/division, positive trigger, positive slope, external trigger on TP31. (E) A channel-2 waveform at TP31 with the following control settings: 2V/division, 10 μs/division, positive trigger, positive slope, trigger on channel 1. (F) A channel-2 waveform at TP32 with the following control settings: 2V/division, 20 ms/division, positive trigger, positive slope, trigger on channel 1. (G) A dual-trace waveform at TP34 and TP33. TP34 is connected to channel 1 with the following control settings: 2V/division, positive trigger, positive slope, 20 ms/division, external trigger on TP32. TP33 is connected to channel 2 with 2V/division.

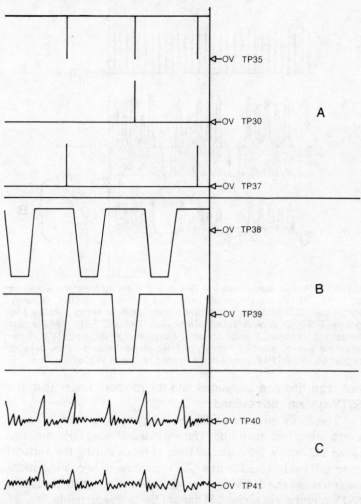

Fig. 6-22. Proper waveforms of memory boards 4, 5, and 6 with an SSTV signal applied. (A) Simulated triple-trace waveform at TP35, TP36, and TP37. In order to simulate this waveform it is necessary to move channel 2 from TP36 to TP37. TP35 is connected to channel 1 with the following control settings: 2V/division, positive trigger, positive slope, 20 ms/division, trigger on channel 2. TP36 is connected to channel 2 with 2V/divison. Then move channel 2 to TP37 with 2V/division. (B) A dual-trace waveform at TP38 and TP39. TP38 is connected to channel 1 with the following control settings: 5V/division, alternate sweep, negative trigger, positive slope, 100 ms/division, trigger on channel 1. TP39 is connected to channel 2 with 5V/division (C) A dual-trace waveform (AC coupled) at TP40 and TP41. TP40 is connected to channel 1 with the following control settings: 0.5V/division, alternate sweep, positive trigger, positive slope, 100 ns/division, trigger on channel 1. TP41 is connected to channel 2 with 0.5V/division.

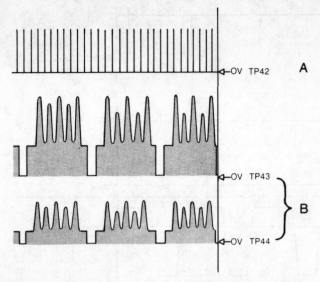

Fig. 6-23. Proper waveforms on D-A board 8. (A) single-trace waveform at TP42. TP42 is connected with the following control settings: 2V/division, 0.2 ms/division, positive trigger, positive slope, external trigger on TP26. (B) A dual-trace waveform at TP43 and TP44. TP43 is connected to channel 1 with the following control settings: 1V/division, alternate sweep, positive trigger, positive slope, 20 μs/division, external trigger on TP20. TP44 is connected to channel 2 with 1V/division.

fast-scan line can be loaded into the memory faster than the SSTV system can respond.

The SSTV picture of 120 or 128 lines is formed by using every other fast-scan line. The normal fast-scan field contains about 242 active lines and 20 lines of black during the vertical sync interval. The 128-line SSTV picture is becoming more common; so the MXV-200 was designed to generate a 128-line SSTV picture, requiring 256 lines of the fast-scan field.

The MXV-200 generates a 120-line picture with 8 lines of gray scale to complete the SSTV field of 128 lines. The eight lines of gray scale at the bottom of the SSTV field allows for easier alignment and contrast and brightness setup.

Video Amplifier

Video from the fast-scan source is applied to video level control R1, then it is fed to emitter-follower Q1 (Fig. 6-25). Q2 is a common-base amplifier with a voltage gain of about 21 dB. The output of Q2 is direct coupled to emitter-follower Q3. One output from Q3 drives the sync separator Q6, Q7, and D1. The sync separator is of the diode-feedback type, which performs very well with the simple sync waveforms from low-cost TV cameras. The sync output from Q7 is coupled to keyed-clamp Q4 and to level translator Q9 to convert the sync to TTL compatible levels. Keyed-clamp Q4 clamps the sync tip of black level control R2. The voltage set by Q8 to present a low-impedance voltage reference to the emitter of Q4. The keyed clamp eliminates the change in video black level with video scene changes. The DC restored video signal is buffered by Q5 to provide a low-impedance drive to the A-D converter. See Fig. 6-26.

Sync Separator

The composite sync from sync separator Q6, Q7, and D1 (Fig. 6-25) is converted to a TTL logic level by Q9 (Fig. 6-27) and triggers one-shot multivibrator U3. U3 is timed to generate a pulse of approximately 45 μs to eliminate any equalizing pulses which occur at 31.5 μs intervals during the vertical sync time. Thus, the output of U3 contains only line-rate pulses occuring every 63.5 μs. See Fig. 6-28. A second output from Q9 is fed to a low-pass filter to remove line-rate sync pulses and leave vertical sync pulses, which are reshaped to TTL logic level by Q10. The output of Q10 is coupled to

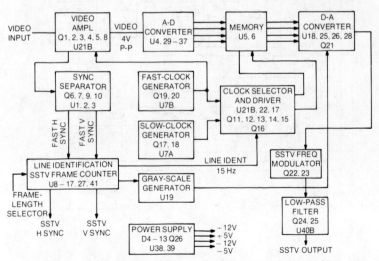

Fig. 6-24. Block diagram of the MXV-200 scan converter. The MXV-200 generates a 120-line picture with 8 lines of gray scale to complete the 128-line field.

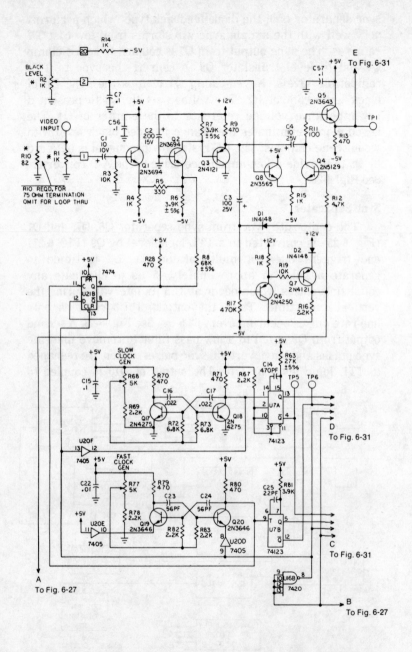

Fig. 6-25. Schematic diagram of the video amplifier, sync separator, and clock generator sections of the MXV-200. Remaining portions of the entire schematic are on following pages.

one-shot multivibrator U1. U1 is used to delay the vertical sync pulse until the end of the vertical blanking period. The first horizontal sync pulse following the end of the pulse from U1 is the first line at the beginning of the SSTV picture. If U1 did not delay past the vertical blanking time, then the first lines in the SSTV picture would be black and contain no picture information. The pulse duration of U1 may be lengthened when used with 625-line 50 Hz standards. Since there are many extra video lines in the fast-scan field, the lengthening of the pulse from U1 will develop an SSTV picture that can be converted from the central portion of the 625-line raster instead of nearer the top. The pulse from U1 is reshaped to a shorter pulse similar to the original vertical sync by U2 and coupled to the line identification circuits. (See Fig. 6-29.)

Line Identification

The function of the line identification circuit (Fig. 6-27) is to generate the SSTV horizontal and vertical sync and to control when to load the fast-scan information. The delayed vertical sync pulse from U2 is divided by counter U8. U8 is programmed by two jumpers with J1 installed and J2 omitted. U8 divides the vertical sync by 4 to obtain a 15 Hz pulse. For 50 Hz standards, install J2 and omit J1 to change U8 to divide by 3, providing the 16.67 Hz required in the 625-line 50 Hz SSTV standards. The output of U8 drives two 4-stage binary counters

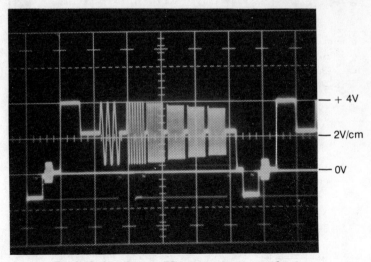

Fig. 6-26. Waveform at TP1 with sweep set to 10 µs/cm.

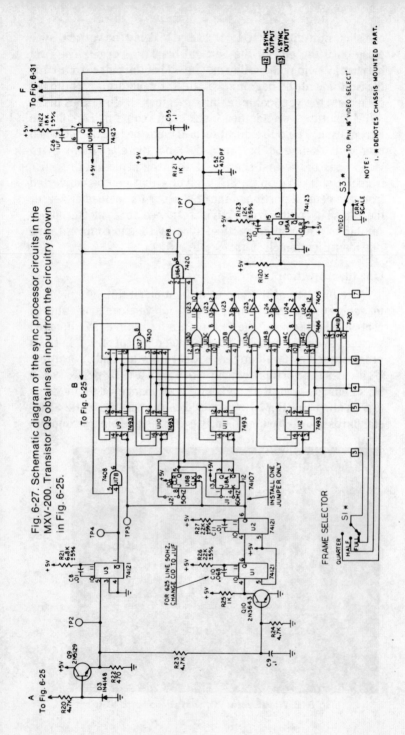

Fig. 6-27. Schematic diagram of the sync processor circuits in the MXV-200. Transistor Q9 obtains an input from the circuitry shown in Fig. 6-25.

U11 and U12. The D output of U12 is connected to the reset lines. This causes both U11 and U12 to be reset to zero when the counters reach a count of 128. The D output of U12 only stays high for about 40 ns once every 8.5 seconds; thus, it cannot be seen on an oscilloscope. This pulse is used to trigger the SSTV vertical sync one-shot multivibrator U14. When the frame-length selector switch S1 is set at one-half, U11 and U12 are reset at count 64, causing a SSTV picture of 64 lines; setting S1 to quarter-frame causes an SSTV picture of only 32 lines. The lead length of these connections to the frame selector should be kept as short as possible and not over 12 inches because of the narrow reset pulses. Seven outputs from U11 and U12 are connected to exclusive NOR gates U13 and U14. A second dual 4-stage binary counter U9 and U10 is driven by the fast-scan horizontal sync. The eight outputs are connected to NAND gate U27. When counters U9 and U10 reach a full count of 255, the output of U27 switches low and causes U17B to shut off further horizontal sync pulses to U9 and U10. When a fast-scan vertical sync pulse occurs, U9 and U10 are reset to zero causing U17B to pass horizontal sync pulses to counter U9. Seven outputs from U9 and U10 are also connected to exclusive NOR gates U13 and U14. Exclusive NOR gates U13 and U14 and inverters U23 and U24 are connected in such a way that the output at TP7 will stay low until counter U9 and

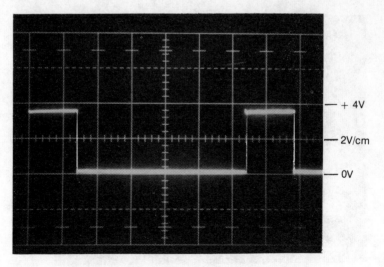

Fig. 6-28. Waveform at TP4 with sweep set to 10 μs/cm.

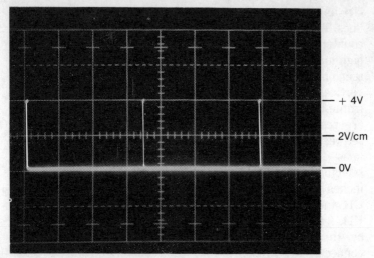

Fig. 6-29. Waveform at TP3 with sweep set at 5 ms/cm.

U10 has the same count as counter U11 and U12. This pulse has a duration of two fast-scan horizontal lines.

The pulse at TP7 (see Fig. 6-30) is combined with pulses from U8 and U9 to generate a pulse at TP8 equal to one fast-scan horizontal line duration and occuring once every four fast-scan fields (three fields for 50 Hz standards). This pulse is used to control when fast-scan information is to be loaded; also, the SSTV horizontal sync is triggered from this pulse.

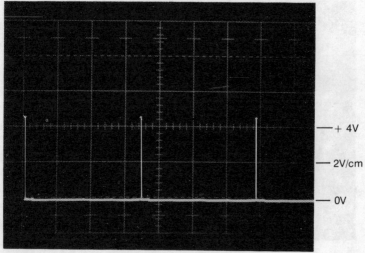

Fig. 6-30. Pulse waveform at TP7 with sweep set to 5 ms/cm.

A-D Converter

The purpose of the A-D converter (Fig. 6-31) is to change the varying voltage level of the video signal into a 4-bit binary code that will represent video level. A 4-bit code resolves the video amplitude into 16 different levels. To reduce the effects of having only 16 levels, a signal is added to the fast-scan video signal before reaching the A-D converter. The dynamic range of the A-D converter is set at 0V for a black level and +4V for a peak white level.

Fifteen comparators are used to convert the video to digital code. Dual comparators of the 711 type were selected due to their availability and cost. The outputs of the comparators are connected together inside the IC. This prevents a standard binary code from being used, but a gray code is used. By connecting the comparators inputs in a particular way, a simple high-speed A-D converter was developed. The number of gates needed to encode the eight outputs into a 4-bit gray code is thus greatly reduced. Voltage reference for each comparator is developed from a voltage divider using a set of 20Ω resistors matched to 2% or using 1% metal film resistors. If an accurate ohmmeter is available, standard 5% carbon resistors, matched within $\pm 2\%$, will work with no loss in accuracy or performance.

Memory

The digital memory (Fig. 6-31) is a dynamic MOS shift register arranged 4 bits wide by 256 bits long. Upon command from the line identification circuits, the memory clock is switched to approximately 5 MHz for a period equal to one fast-scan horizontal line. This loads the memory with fast-scan video. The clock frequency is higher than needed to load 256 bits; however, since the memory will only store 256 bits, the fast-scan sync and blanking information are lost. At the end of the fast-scan line, the clock is switched back to the slow-scan rate of 2500 Hz. This clock frequency will allow only about 230 bits to be clocked from the memory during each slow-scan horizontal line. This process of loading greater than 256 lines and only unloading 230 bits performs the conversion of the 4-by-3 rectangular fast-scan TV standards to the 1-by-1 square SSTV standards. The output from memory U5 is coupled to quad-exclusive NOR gate U6. IC U6 is a gray-code-to-binary-code converter because the D-A converter will not work

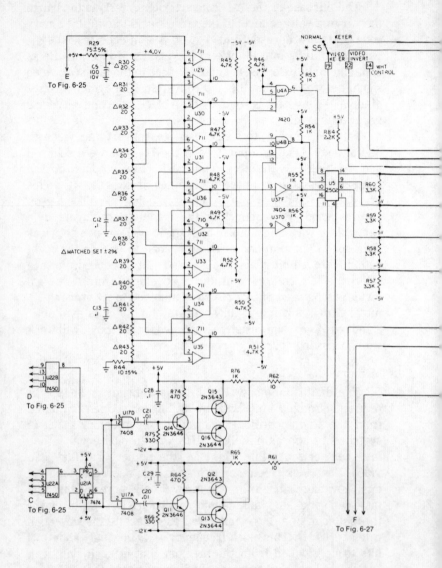

directly on gray code. Video-invert control S2 is connected to U6C to invert the polarity of the MSR. This in turn inverts the polarity of the binary-code output of U6. Thus, video-invert control S2 is a TTL logic level which may be controlled by a front panel switch or from other TTL logic to generate special

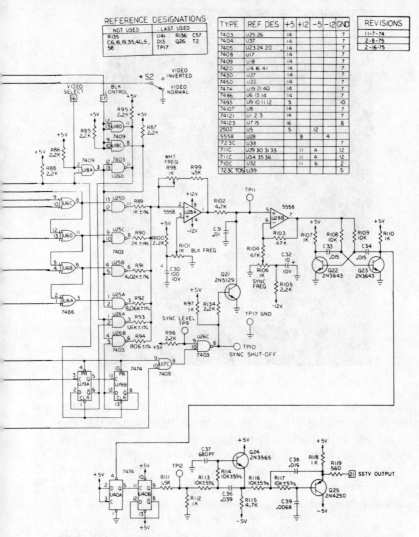

Fig. 6-31. Schematic diagram of the A-D converter, memory, D-A converter, clock selector and driver, gray-scale generator, SSTV frequency modulator, and low-pass filter sections of the MXV-200.

effects. The MSB is brought out to use as a special-effects keying signal, as illustrated in Figs. 6-31, 6-32, and 6-33.

Clock Generator

The fast-clock generator (Fig. 6-25) consists of gated multivibrator Q19 and Q20. Q19 and Q20 form a conventional

Fig. 6-32. A special-effects raster instrumented by the MSB.

astable multivibrator with adjustable base voltage to set the correct operating frequency of approximately 5 MHz. The fast-clock oscillator is only gated *on* during the 15 Hz 63 μs line identification pulse. Due to this low duty cycle, it is difficult to see the output signal on an oscilloscope. To enable the oscillator to run continuously, U16 may be removed to allow observation on an oscilloscope waveform as shown in Fig. 6-34 at TP6. The switching characteristic of Q19 and Q20 are important due to the high frequency, and transistor substitution is not recommended. The oscillator output is reshaped by U7B into a pulse of approximately 100 ns.

A second output from the fast-clock oscillator drives U21B, a divide-by-two counter. The 2.5 MHz square-wave signal from U21B is added to the fast-scan video amplifier at an amplitude equal to 1/16 of the full-scale input to the A-D converter. This signal added to the video—equal to one LSB—will greatly improve the gray-scale rendition of the A-D converter with only a small reduction in signal-to-noise ratio.

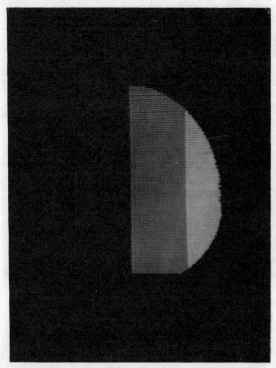

Fig. 6-33. Another special-effects raster brought about by the MSB.

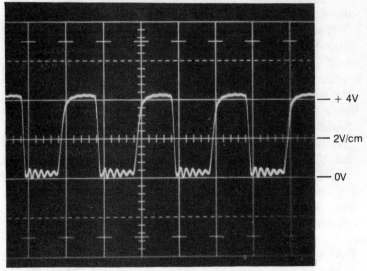

Fig. 6-34. Waveform of fast-clock generator output at TP6 with U16 removed for a continuous output. Sweep is set at 0.1 µs/cm.

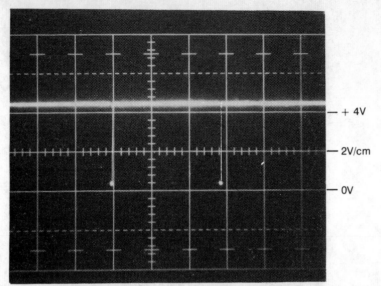

Fig. 6-35. Waveform of slow-clock generator output at TP5. Sweep is set at 100 μs/cm.

The slow-clock generator (Fig. 6-25) is identical to the fast-clock oscillator, except that some component value changes must be made to run at 3500 Hz. The clock runs continuously, but is reset by the 15 Hz line identification pulse to synchronize the oscillator, reducing horizontal jitter in the output picture.

Slow-clock oscillator Q17 and Q18 is reshaped into narrow pulses by U7A as shown in Fig. 6-35 at TP5. The outputs of both clock generators are coupled to selector U22 (Fig. 6-31). During the 63 μs line identification pulse, the fast clock is selected; during the remaining time, the slow clock is used. The output of the clock selector is coupled to U12A, which divides the output by two, and then it is coupled to U17A and U17D to remove alternate clock pulses. The outputs of U17A and U17D are 180° out of phase. These two signals are amplified to a level of +5V to −12V to drive the MOS shift-register-memory U5. Each clock driver consists of pulse amplifiers Q11 and Q14, followed by a complimentary emitter-follower buffer Q12, Q13, Q15, and Q16.

D-A CONVERTER

The 4-bit binary-coded SSTV video from U6 is coupled to U25 (Fig. 6-31), which is part of the D-A converter. Additional control circuits are included to simplify alignment and to create some special effects. The D-A conversion is accomplished by controlling the gain of U28A by a series of binary-weighted resistors. A DC voltage is coupled to pin 3 of U28A for amplification. The degree of amplification is a function of the resistor connected to pin 2 of U28A. U25 and U26 are open-collector NAND gates. When either input to the NAND gate is low, the output is an open circuit. Thus, if a low level is presented to U25A, U25B, U25C, U25D, U26A, and U26B, all six gates will have an open-circuit output effectively disconnecting R89 through R94 from ground. Therefore, U28A will be operating as a unity-gain voltage follower, and the output voltage at pin 1 will be equal to the input voltage at pin 3. This voltage corresponds to a black level in video. When any of the NAND gates have a high level on both inputs, then the resistor at each NAND gate output will be connected to ground; this will increase the gain of U28A. It follows that the output level of U28A is proportional to the binary code presented to the inputs of U25.

The gray scale is generated by dividing the 60 Hz pulse at TP7 by four with U19 and driving NAND gates U26A and U26B. The divider is reset by the SSTV horizontal sync pulse to insure that the gray scale starts with black at the left side of the picture. The SSTV horizontal and vertical sync are combined in U17C, inverted by U26C, and drive Q21 to add the composite SSTV sync information to the SSTV video output from U28A. The complete SSTV video signal at TP11 (Fig. 6-36) is buffered by U28B; a DC offset is added to set the sync voltage for the FM modulator.

SSTV Frequency Modulator and Filter

The composite SSTV video signal from U28B is used to vary the base voltage on astable multivibrator Q22 and Q23 (Fig. 6-31). This form of frequency modulator has the advantage of excellent linearity and low cost. The output of the multivibrator is reshaped by U40A to improve the signal rise time. The frequency modulator runs at twice the frequency required and is divided by two to the correct frequency by U40B. Thus, the output of U40B is exactly a 50% duty-cycle square wave containing only odd harmonics. The elimination of even harmonics reduces the filter selectivity requirements. The low-pass filter is 4-pole Butterworth active filter Q24 and Q25.

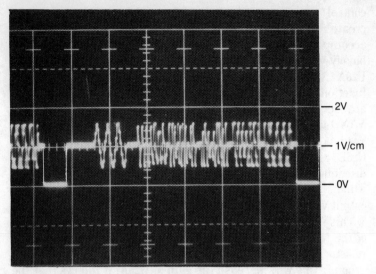

Fig. 6-36. Waveform of SSTV signal at TP11. Sweep is set at 10 ms/cm.

Power Supply

The power supply (Fig. 6-37) is not special, and the voltages are not critical. The +5V supply should be within 5% to insure proper operation of the TTL logic. In order to keep costs down and to use parts that are common, 12.6V transformers were used. The load is equal on each transformer, allowing the primary to be connected in parallel for 110V operation and in series for 220V operation. These transformers should be of the same type to insure balanced operation. The current requirements on the −5V and −12V supplies is low enough to need only simple zener regulators. The +12V supply load is less than 100 mA, so a common 723 regulator was chosen. Regulator U39 should be in a TO-100 metal can with a small-finned heat sink attached. The +5V regulator, U38, is a conventional circuit, but the power-pass transistor is inverted to allow it to regulate down to within 1V above the 5V output. This improves the regulator under low line-voltage operation. Most 3-terminal regulators need an input 3V above the output to operate. The lower voltage drop across Q26 also reduces the heating; a small aluminum heat sink is mounted on Q26.

Construction

The MXV-200 has been designed to operate with a PC board. The board designed is a double-sided board. On the

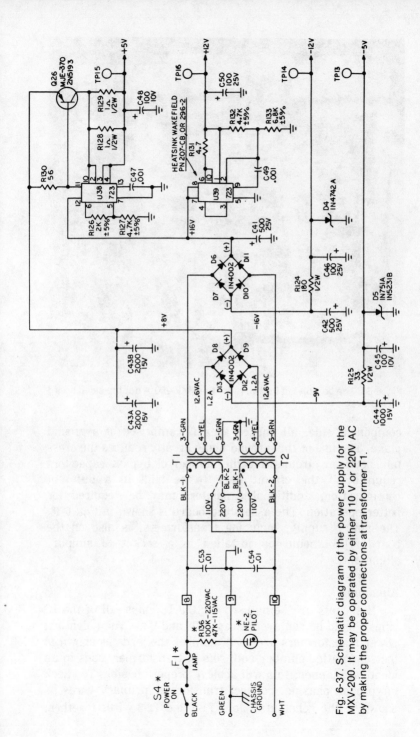

Fig. 6-37. Schematic diagram of the power supply for the MXV-200. It may be operated by either 110 V or 220V AC by making the proper connections at transformer T1.

Fig. 6-38. The completed PC board for the MXV-200. A double-sided board is employed.

component side, all unused area is comprised of a ground plane. The use of this ground plane greatly reduces the cross talk between circuits and the number of bypass capacitors required; if the circuit were to be built in a different configuration, additional capacitors may be required for better operation. The assembled board is shown in Fig. 6-38. The power supply, including transformers, is also on the board. U16 should be installed in a socket to simplify alignment.

Alignment

If sockets are being used for the IC chips, all of the IC chips should be removed except U38 and U39. When identical power transformers are used, number the leads as shown in the schematic; number both sets of transformer leads in an identical manner; this will achieve proper phasing. To check for proper phasing, connect transformer primary wires as shown in the schematic. Join T1-5 and T2-3 wires together;

with an AC voltmeter, measure the voltage from T1-3 to T2-5. This reading should be equal to twice the voltage of one transformer. If a low or zero reading is achieved, reverse the primary wires of one transformer; the proper voltage should now be obtained. With proper phasing, the transformer leads should be numbered as shown in the schematic because the board has the same number designations. Before applying power, connect a voltmeter to TP15. Apply power; the voltage at TP15 should not exceed +5.3V because damage to the TTL IC chips may result. Check remaining power supply voltages at TP13, TP14, and TP16; all voltages should be within 10% of the required voltages. Insert all of the IC chips; check for the proper voltages.

Connect a fast-scan video source to the input connector. Many cameras require a 75Ω termination to operate properly. The MXV-200 has an impedance of 1000Ω. If R10 is added as shown in the schematic, the equivalent impedance will be 75Ω. Set R1 and R2 to midrange. This should give a video level at TP1 of approximately 5V from sync tip to peak white level (P-P); see Fig. 6-34. Remove U16; this will enable fast-clock oscillator Q19 and Q20 to run continuously. Adjust R77 for a frequency at TP6 (Fig. 6-25) of 5 MHz. Reinstall U16; the fast-clock signal will only run for a period of 63 μs every 64 ms. Adjust R68 for a slow-clock frequency of 3500 Hz at TP5 (Fig. 6-35).

The remaining alignment requires setting the operating frequency of the FM modulator. Operation of the remaining circuits is checked by comparing the waveforms against the typical waveforms shown in this section and referring to the operating theory in previous paragraphs.

The alignment of the SSTV frequency modulator is simple, and if performed in this sequence, has no interaction between adjustments. The alignment is best achieved by using a frequency counter or calibrated oscilloscope. The FM oscillator runs at twice the output frequency of normal SSTV. Connect frequency counter to TP12. Ground TP9, and adjust R106 for a frequency of 1200 Hz at TP12. Remove ground from TP9, and ground TP10. Ground the lead marked BLACK CONTROL at board connector pin 17. Adjust R101 for 1500 Hz at TP12. Leave TP10 grounded, and remove ground from connector pin 17, and then ground lead marked WHITE CONTROL at connector pin 14. Adjust R98 for 2250—2300 Hz at

TP12. In some cases, it may be necessary to change the value of R99 to get R98 into this frequency range; larger values of R99 increase the frequency. Remove all added ground wires at TP9, TP10, and connector pin 17 or 14. At this time, a SSTV picture should appear on the monitor. With video selected, the picture should occupy the full width of the SSTV monitor. Missing picture or noise on the left side of the picture indicates that the fast-clock frequency is not high enough; adjusting R77 will cause the picture to move to the left. Adjusting slow-clock frequency control R68 causes the right side of the picture to expand. By adjusting R77 and R68, a normal size picture should be achieved with the proper 1 by 1 aspect ratio.

For normal contrast picture, R2 should be adjusted for the black portion of the picture to be at ground level at TP1. Full white level may be set by R1 to +4V at TP1. To increase the contrast in certain portions of the picture, the white level may be increased above +4V; and the black level may extend below ground potential. The MXV-200 will not operate or generate a gray scale unless a fast-scan video source is connected to the input. Fast-scan sync pulses are also required for operation.

Applications

The MXV-200 may be used with a single fast-scan source in the normal manner; or, by utilizing some of the external control lines provided, some special effects can be created. When wired as shown, the last eight lines of the SSTV picture contain the four-step gray scale. This gray scale can be removed by disconnecting the lead from board connector pin 7 to the video selector switch S3.

The MXV-200 is capable of some special effects using no additional equipment except a fast-scan source. By using a fast-scan camera aimed at white lettering or objects, the special effects shown in Fig. 6-32 may be generated. Video level R1 should be adjusted so that the information to be inserted into the picture extends above +2V at TP1. In Fig. 6-33, the fast-scan camera is aimed at a white circle, S5 is set for KEYER, S3 is set for GRAY SCALE, and S2 is set for NORMAL.

By utilizing the SSTV horizontal and vertical sync outputs from the MXV-200 to lock a source such as a flying-spot

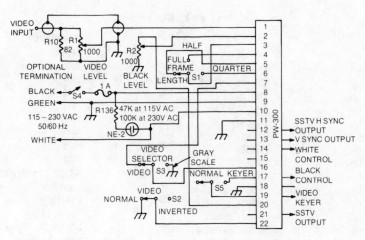

Fig. 6-39. Chassis wiring diagram for the MXV-200.

scanner or keyboard, many additional effects can be generated using the BLACK CONTROL, WHITE CONTROL, VIDEO INVERT, and VIDEO SELECT TTL logic inputs (Fig. 6-39). Any SSTV sources used with the MXV-200 must be locked to the same sync pulses. The SSTV video used for special effects cannot be FM; but must be converted to TTL logic levels by a suitable comparator, such as a LM710 or equivalent.

PC boards and parts kits are available from the designer, W6MXV, Michael Tallent, 6941 Lenwood Way, San Jose, California 95120. See Fig. 6-40.

Fig. 6-40. The completed MXV-200. PC boards and parts kits may be ordered from the designer. W6MXV.

MXV-200 PARTS LIST

Part	Description
Q1	2N3694. 2N3904
Q2	2N3694. 2N3904
Q3	2N4121. 2N4122
Q4	2N5129
Q5	2N3643. 2N697
Q6	2N4250
Q7	2N4121. 2N4122
Q8	2N3565
Q9	2N5129
Q10	2N3643
Q11	2N3646. MPS3646
Q12	2N3643. 2N3642
Q13	2N3644. 2N3645
Q14	2N3646. MPS3646
Q15	2N3643. 2N3642
Q16	2N3644. 2N3645
17	2N4275. 2N3646
Q18	2N4275. 2N3646
Q19	2N3646. 2N708. 2N2369
Q20	2N3646. 2N708. 2N2369
Q21	2N6129
Q22	2N3643. 2N3642
Q23	2N3643. 2N3642
Q24	2N3565
Q25	2N4250
Q26	MJE-370. 2N6193
U1	74121
U2	74121
U3	74121
U4	7420. 7440
U5	MM1402AN (N). 2502B (sig)
U6	7486
U7	74123
U8	74107
U9	7493
U10	7493
U11	7493
U12	7493
U13	7486
Ux4	7486
U15	74123
U16	7420. 7440
U17	7408
U18	7409
U19	7474
U20	7405
U21	7474
U22	7450. 7451
U23	7405
U24	7405
U25	7403
U26	7403
U27	7430
U28	N5558. MC1458 mini DIP
U29	LM711CN or equiv. 14 pin
U30	LM711CN or equiv. 14 pin
U31	LM711CN or equiv. 14 pin
U32	LM710CN or equiv. 14 pin
U33	LM711CN or equiv. 14 pin
U34	LM711CN or equiv. 14 pin
U35	LM711CN or equiv. 14 pin
U36	LM711CN or equiv. 14 pin
U37	7404
U38	LM723CN or equiv 14 pin
U39	LM723CH or equiv. 10 pin TO-100
U40	7474
U41	7420. 7440

Note: all IC chips are DIP pkg except U39.

T1. T2	12.6V AC at 1A center tap Radio Shack #273-1505 or equiv.
TP1–17	test point terminals H. H. Smith #2024 or equiv.
Wr1	18-inch length two conductor shielded cable Belden #8451

MXV-200 PARTS LIST (cont.)

Part	Description
S01−4	1-inch threaded standoff 6-32 threads. H. H. Smith 8427. Used to mount T1, T2
S1	Frame length selector, 1-pole 3-position
S2	Video invert SPDT rocker switch
S3	Video selector SPDT rocker switch
S4	Power SPST 3A rocker switch
S5	Video keyer SPDT rocker switch
F1	1A fuse and holder AGC-1
J1	22-pin connector Amphenol 143-022-01
J2-J7	Phono jacks, Switchcraft 3501FP or equiv.
R1	1000Ω 2W pot (chassis)
R2	1000Ω 2W pot (chassis)
R3	10K 1/4W
R5	330Ω 1/4W
R6	3900Ω 1/4W 5%
R7	3900Ω 1/4W 5%
R8	68K 1/4W 5%
R9	470Ω 1/4W
R10	82Ω 1/4W optional
R11	100Ω 1/4W
R12	4700Ω 1/4W
R13	470Ω 1/4W
R14	1000Ω 1/4W
R15	1000Ω 1/4W
R16	2200Ω 1/4W
R17	470K 1/4W
R18	1000Ω 1/4W
R19	10K 1/4W
R20	4700Ω 1/4W
R21	6800Ω 1/4W 5%
R22	470Ω 1/4W
R23	4700Ω 1/4W
R24	4700Ω 1/4W
R25	1000Ω 1/4W
R26	22K 1/4W 5%
R27	22K 1/4W 5%
R28	470Ω 1/4W
R29	75Ω 1/4W 5%
R30	20Ω 1/4W 1%
R31	20Ω 1/4W 1%
R32	20Ω 1/4W 1%
R33	20Ω 1/4W 1%
R34	20Ω 1/4W 1%
R35	20Ω 1/4W 1%
R36	20Ω 1/4W 1%
R37	20Ω 1/4W 1%
R38	20Ω 1/4W 1%
R39	20Ω 1/4W 1%
R40	20Ω 1/4W 1%
R41	20Ω 1/4W 1%
R42	20Ω 1/4W 1%
R43	20Ω 1/4W 1%
R44	10Ω 1/4W 5%
R45	4700Ω 1/4W
R46	4700Ω 1/4W
R47	4700Ω 1/4W
R48	4700Ω 1/4W
R49	7400Ω 1/4W
R50	4700Ω 1/4W
R51	4700Ω 1/4W
R52	4700Ω 1/4W
R53	1000Ω 1/4W
R54	1000Ω 1/4W
R55	1000Ω 1/4W
R56	1000Ω 1/4W
R57	3300Ω 1/4W
R58	3300Ω 1/4W
R59	3300Ω 1/4W
R60	3300Ω 1/4W
R61	10Ω 1/4W
R62	10Ω 1/4W
R63	27K 1/4W 5%
R64	470Ω 1/4W
R65	1000Ω 1/4W
R66	330Ω 1/4W
R67	2200Ω 1/4W

MXV-200 PARTS LIST (cont.)

Part	Description
R68	5000Ω pot IRC-X201-R502B
R69	2200Ω 1/4W
R70	470Ω 1/4W
R71	470Ω 1/4W
R72	6800Ω 1/4W
R73	6800Ω 1/4W
R74	470Ω 1/4W
R75	330Ω 1/4W
R76	1000Ω 1/4W
R77	5000Ω pot IRC-X201-R502B
R78	2200Ω 1/4W
R79	470Ω 1/4W
R80	470Ω 1/4W
R81	3900Ω 1/4W 5%
R82	2200Ω 1/4W
R83	2200Ω 1/4W
R84	2200Ω 1/4W
R85	2200Ω 1/4W
R86	2200Ω 1/4W
R87	2200Ω 1/4W
R88	2200Ω 1/4W
R89	1000Ω 1/8W 1%
R90	2000Ω 1/8W 1%
R91	4020Ω 1/8W 1%
R92	8060Ω 1/8W 1%
R93	1620Ω 1/8W 1%
R94	806Ω 1/8W 1%
R95	2200Ω 1/4W
R96	2200Ω 1/4W
R97	1000Ω 1/4W
R98	1000Ω pot IRC-X201-R102B
R99	1500Ω 1/4W
R100	2200Ω 1/4W
R101	1000Ω pot IRC-X201-R102B
R102	4700Ω 1/4W
R103	47K 1/4W
R104	47K 1/4W
R105	2200Ω 1/4W
R106	1000Ω pot IRC-X201-R102B
R107	1000Ω 1/4W
R108	10K 1/4W
R109	10K 1/4W
R110	1000Ω 1/4W
R111	1500Ω 1/4W
R112	1000Ω 1/4W
R113	10K 1/4W 5%
R114	10K 1/4W 5%
R115	4700Ω 1/4W
R116	10K 1/4W 5%
R117	10K 1/4W 5%
R118	1000Ω 1/4W
R119	560Ω 1/4W
R120	1000Ω 1/4W
R121	1000Ω 1/4W
R122	18K 1/4W 5%
R123	12K 1/4W 5%
R124	180Ω 1/2W
R125	33Ω 1/2W
R126	2000Ω 1/4W 5%
R127	4700Ω 1/4W 5%
R128	1Ω 1/2W or 1/4W
R129	1Ω 1/2W or 1/4W
R130	56Ω 1/4W
R131	4.7Ω 1/4W
R132	4700Ω 1/4W 5%
R133	6800Ω 1/4W 5%
R134	2200Ω 1/4W
R135	not used
R136	47K (115V AC power) 100K (230V AC power)
C1	10 μF 10V
C2	200 μF 15V
C3	100 μF 25V
C4	10 μF 15V
C5	100 μF 10V
C6	not used
C7	0.1 μF disc 10V
C8	0.01 μF 10% Mylar
C9	0.1 μF 10% Mylar
C10	0.068 μF 10% Mylar
C11	0.01 μF 10% Mylar
C12	0.1 μF disc 10V
C13	0.1 μF disc 10V

MXV-200 PARTS LIST (cont.)

Part	Description	Part	Description
C14	470 pF 5% mica DM-15	C38	0.015 μF 10% Mylar 100V
C15	0.1 μF disc 10V	C39	0.0068 μF 10% Mylar 100V
C16	0.022 μF 10% Mylar	C40	not used
C17	0.022 μF 10% Mylar	C41	500 μF 25V
		C42	500 μF 25V
C18	not used	C43A, B	2000 μF 15V
C19	not used	C44	1000 μF 15V
C20	0.01 μF disc 100V	C45	100 μF 10V
C21	0.01 μF disc 100V	C46	100 μF 25V
C22	0.01 μF disc 100V	C47	0.001 μF disc 100V
C23	56 pF 5% mica DM-15	C48	100 μF 10V
C24	56 pF 5% mica DM-15	C49	0.001 μF disc 100V
C25	22 pF 5% mica DM-15	C50	100 μF 25V
		C51	not used
C26	1 μF 20V Tant Sprague 196D	52	470 pF 10% disc or mica
C27	10 μF 20V Tant. Sprague 196D	C53	0.01 μF 1000V disc
C28	0.1 μF disc 10V	C54	0.01 μF 1000V disc
C29	0.1 μF disc 10V		
C30	100 μF 10V	C55	0.1 μF 10V
C31	0.01 μF disc 100V	C56	0.1 μF 10V disc
C32	10 μF 10V	C57	0.1 μF 10V disc
C33	0.015 μF 10% Mylar 100V	D1	1N4148, 1N914
		D2	1N4148, 1N914
C34	0.015 μF 10% Mylar 100V	D3	1N4148, 1N914
		D4	1N4742A 12V 5% 1W
C35	not used		
C36	0.039 μF 10% Mylar 100V	D5	1N751A, 1N5231B 5.1V 5%
C37	680 pF 5% mica DM-15	D6–13	1N4002 100 PIV 1A

W9NTP SLOW-SCAN COLOR CONVERTER

In order to understand how a slow-scan color converter works, it will be necessary to review the basic principles of color TV.

It has been recognized for many decades that color TV pictures can be displayed by using only three basic colors to produce all of the hues that are desirable in the displayed picture. Much has been written on this subject, and color TV is

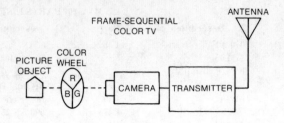

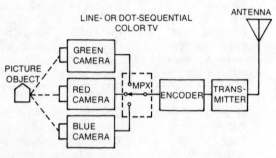

Fig. 6-41. Two basic methods—frame sequential and line or dot sequential—of transmitting colors separately.

with us today because of all of this basic work. (Those of you interested in reading more on the subject should consult *Fink's Television Handbook*.)

Based on this principle color TV can be transmitted provided red, blue, and green pictures are transmitted separately, then recomposed at the receiver. Industry has investigated two basic methods of doing this. Each of the methods are shown in Fig. 6-41.

The first method is the famous CBS frame-sequential system; the second method is the RCA dot-sequential method. Both of these systems are capable of generating high-quality color TV pictures.

The first method was doomed to failure because of anticipated mechanical viewing problems. The camera wheel could have been eliminated by using three switched cameras, but the viewing problem of rotating wheels or drums was never satisfactorily solved.

The designers of color TV also had to solve a tough RF-spectrum problem. Monochrome TV uses a 6 MHz channel allocation, and it was felt that it was necessary to put the three color signals into this same size spectrum. In addition it was also necessary to make the color signal compatible with all existing monochrome sets.

It has been known for a long time that frequencies that are related to the horizontal-line rate, by being an odd multiple of half the horizontal line rate, would produce very little beats in the picture. The decision to put two of the three color channels on a subcarrier of 3.58 MHz is based on this mathematical principle. In order to get two independent channels on a single carrier, it is necessary to use the same principles used in phasing-type SSB transmitters. This means that two balanced modulators are used that permit orthogonal (noninterfering) signals on a single carrier. In the case of SSB, the phasing of the signals is arranged to cancel out one sideband. In the case of color TV, these two signals will be fully independent DSB signals placed at the high end of the 6 MHz allocation. The spectrum is shown below in Fig. 6-42.

Through many years of subjective testing, it was found that it is not necessary to transmit the full bandwidth for each of the three basic colors, red, green, and blue. The three channels that are transmitted differ greatly in bandwidth, and this makes it possible to squeeze three channels into the 6 MHz monochrome bandwidth. The three signals are the luminance channel and two chrominance channels of lesser bandwidth. The wider bandwidth of the two chrominance channel is 1.5 MHz and is called the I or in-phase channel. It is used to transmit orange or cyan changes in color. The other chrominance channel of 0.5 MHz bandwidth is called the Q

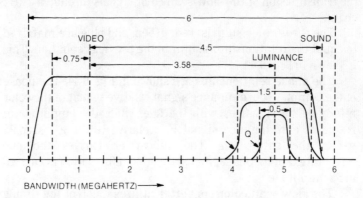

Fig. 6-42. Frequency spectrum of color TV. Bandwidths are as follows: composite signal, 6 MHz; video – sound, 4.5 MHz; video – luminance, 3.58 MHz; I channel, 1.5 MHz; Q channel, 0.5 MHz.

channel or quadrature channel. This channel carries changes in the color from white to green and magenta.

The mathematical relationship between the various channels and color signals is given by the following equations:

$$E_I = 0.6\,E_R - 0.28\,E_G - 0.32\,E_B$$
$$E_Q = 0.21\,E_R - 0.52\,E_G + 0.31\,E_B$$
$$E_Y = 0.3\,E_R + 0.59\,E_G + 0.11\,E_B$$

From which it can be shown that

$$E_I = 0.74\,(E_R - E_Y) - 0.27\,(E_B - E_Y)$$
$$E_Q = 0.48\,(E_R - E_Y) + 0.41\,(E_B - E_Y)$$

where:

E_R = red voltage
E_G = green voltage
E_B = blue voltage
E_I = in-phase voltage
E_Q = quadrature voltage
E_Y = luminance voltage

There is too much theory concerning these equations to be discussed here. These are the basic color equations and the ones that will be used in the development of a slow-scan color converter.

In the design of a color converter, there are two places where the above design criteria will be considered: The first is the display of the memory on the color TV set, and second in the transmission of the slow-scan color signals through a 3 kHz channel.

The three color signals, red, green, and blue are matrixed to equal the relationships shown in the previous equation. This is shown in Fig. 6-43.

The phase relationships are shown in Fig. 6-44. The phase of the complete chrominance signal relative to that of the color reference burst varies with the hue, while the amplitude of resultant chrominance signal E_C relative to the E_Y signals varies the saturation. The subcarrier carries the hue information in its phase, and saturation information is carried in its amplitude.

The slow-scan color converter utilizes much of the timing circuitry that is contained in the W0LMD slow- to fast-scan converter. The basic timing of the converter is used for all three color memory channels.

Each memory channel contains 4 by 16384 bits. Each channel is used to store a monochrome representation of one of the basic colors (R, G, B) of the displayed color picture. These three stored pictures can be derived from a color camera, a color-wheel camera, or from a slow-scan color tape prepared by WB8DQT.

The most interesting way to load the memory is frame sequential. A properly synchronized color filter is rotated in front of the lens at the correct time for memory loading. The fastest that the three memories can be loaded is 3 times each 1/60 second, or 1/20 second.

The memory provides for parallel loading in either fast-scan or slow-scan modes. It will be seen later that when the converter is used for generating a color picture, there are

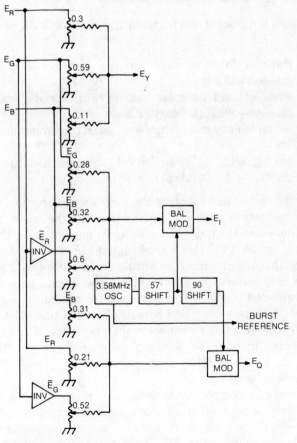

Fig. 6-43. Basic diagram of color matrix and encoder.

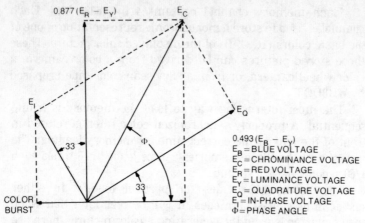

Fig. 6-44. Color-phase vectors.

four options for loading and transmitting color pictures. These are:

- Parallel load by color camera with serial transmission—20 sec
- Parallel load by color camera with parallel transmission—500 Hz subcarrier, 8 sec
- Serial load by color wheel with serial transmission—20 sec
- Serial load by color wheel with parallel transmission—500 Hz subcarrier, 8 sec

A word should be said about the memory size required for color transmission. It is undoubtedly true that the size of the red memory and the blue memory can be much smaller than that of the green. Probably a good estimate of memory usage would be that the red memory should contain 16384 by 2 bits and the blue contain 16384 by 1 bits. Experiments have not been completed yet to define a system for minimum use of memory. An interesting side benefit of having the full 4-bit memories in each color channel is the possibility of running the memory in a high-resolution monochrome mode. This results in a 128-by-384-by-4-bit picture or a pseudo 256-by-384-by-4-bit picture.

In order to completely understand the operation of the color transmission system, consider the block diagram shown in Fig. 6-45. Three color sources are used. If a color camera is available, the multiplexer (MPX) can simultaneously load all

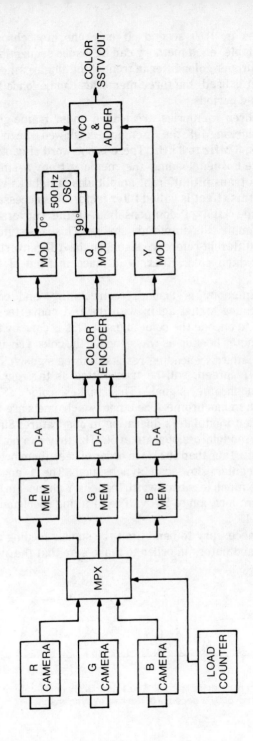

Fig. 6-45. A slow-scan color TV simultaneous transmission system.

three memories in 1/60 second. If only one monochrome camera is available, each memory can be loaded sequentially by placing the proper color filter in front of the single camera. If a color wheel is used, the three memories can be loaded in three 1/60-second periods.

Once the three memories are loaded under frame-grab control, the clocking of the memory is slowed down to slow-scan rates, 3 MHz to 3 kHz. There is one word of caution to be considered when loading the memory from a single camera: filters transmit different amplitudes of light as well as frequency—this effect is called filter factor. It is necessary to provide some kind of compensation of the conversion process to eliminate this amplitude change. This is best done by providing a different reference signal on the A-D converter, depending on which color filter is in place in front of the camera.

Once the memory is loaded, recirculating, and converted to an analog signal again with the D-A converter, it is now possible to encode the color signals. This is done by the same potentiometer hookup as was done for the color viewing. (Fig. 6-43). The matrix encoding results in three signals. The Y signal, usually green, will be transmitted as the regular slow-scan monochrome signal. This makes color SSTV compatible with monochrome. The other two channels are fed into two balanced modulators operating in quadrature. Since these balanced modulators operate at 500 Hz, they can not be modulated any further than 500 Hz in order to keep them below the 1200 Hz regular slow-scan sync signal. The I and Q channels can be much less bandwidth than the Y channel in the normal spectrum location of SSTV. The resulting spectrum is shown in Fig. 6-46.

It may be necessary to band limit the signals feeding the two balanced modulators in order to make sure that the sync

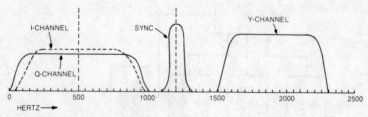

Fig. 6-46. Slow-scan color spectrum. The I and Q channels cannot be modulated more than 500 Hz because of the nearby sync channel.

signal is not interfered with. The spectrum is fed into a HF transmitter through the microphone jack. Monochrome viewers will use only the Y channel. Any interference created by the I and Q channels can be eliminated by using a low-pass filter, 500–1000 Hz, in front of the voltage-controlled oscillator (VCO).

The color receiving system is shown in Fig. 6-47. A conventional receiver is used to tune in an SSTV signal. The output signal is fed into the two filters. One filter passes 0–1000 Hz and the other passes 1000–2500 Hz signals. The filter makes sure that the Y energy and the I and Q energy have the best chance of getting to their respective demodulators. This type of high-pass filter, 1000–2500 Hz, is in common use today by many SSTV operators since it helps eliminate interference coming in below the regular slow-scan signal.

The Y signal is detected and changed to a fast-scan signal by a W0LMD slow- to fast-scan timing circuit. The I and Q signals are fed into the two balanced demodulators with a 500 Hz signal. The 500 Hz signal is phase locked to a burst signal that is sent with the I and Q signals. The length of this burst signal is not definitely determined, but it is possible to allow it to run continuously since it will produce very little interference. Once locked it will produce a steady DC signal which can be filtered out of the picture. If some unforeseen interference does develop, it may be necessary to increase the length of the slow-scan horizontal sync pulse and send a sample of 500 Hz for a few cycles. One period of 500 Hz is 2 ms long, so 5 periods would result in a burst of 10 ms, which is considerably longer than the present horizontal sync pulse.

It is obvious from the diagram that if a frame-sequential slow-scan picture is desired, the two filters, balanced modulators, and 500 Hz oscillator can all be eliminated. This is the system recommended for beginners. The MPX is used to load each color scene in sequence (8 seconds) so that 24 seconds are consumed for loading and transmission of each color picture. Once the color memories are loaded with either sequential or simultaneous information, clocking through the memory is increased from 3 kHz to 3 MHz. Since the signals are already encoded, the memory signals can be fed into the respective fast-scan modulators.

The I signal is fed into a balanced modulator that is translated to 3.58 MHz at 0° phase. The Q channel is fed into a

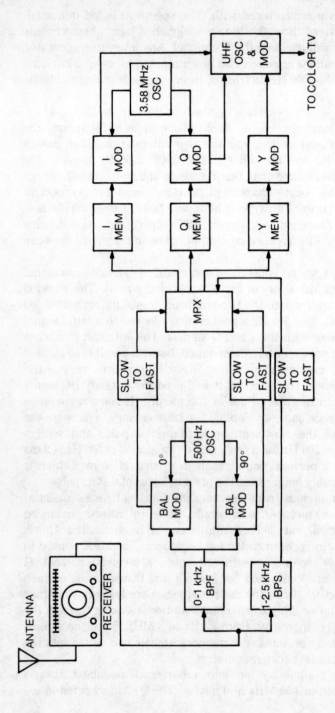

Fig. 6-47. A slow-scan color TV receiving system with an 8-second frame. This system is ideal for beginners.

182

balanced modulator that is supplied with a 3.58 MHz quadrature signal. The Y signal is added to the output of the I and Q balanced modulators to modulate a VHF oscillator that oscillates on an unused TV channel.

It is possible to decode the I, Q, and Y signals to the respective R, G, and B signals, then connect them directly to a color TV set. This eliminates all balanced modulators and VHF oscillators, and does not require the delicate adjustments necessary to get a 3.58 MHz quadrature modulation system working. Some TV sets, such an the Trinitron (Sony), have connections that make it very easy to put video signals directly on the color CRT. Other sets such as Zenith have an internal hookup for the presentation of a color image that is a bit more complicated. The Y or monochrome signal is connected directly to the cathode of the CRT. Usually the sync is derived from this signal also. The three CRT grids are supplied with R-Y, B-Y, and G-Y signals (red minus monochrome, etc.). Since the signals presented to the grid and cathode subtract internally, it is easily understood that the CRT actually sees B, R, and G. This may be a problem for you. It is possible to make up these difference signals very easily, but a better scheme is to supply only a composite sync signal to the Y channel; the R-Y channel is supplied with a red video signal, the B-Y channel is supplied with a blue signal, and the G-Y channel is fed a green signal. In this way the background brightness can be governed with the *brightness control* on the TV set. The normal luminance channel is higher bandwidth than the color circuits; but, in the case of SSTV, this is a benefit since the bandwidth of the converted slow-scan picture is about 2 MHz and looks better when it has some bandwidth limiting. A simple diagram of this direct system of supplying color to the fast-scan monitor is shown in Fig. 6-48.

Basic Circuit Details

Two uses are made of the MC 1596 balanced modulators; one is for generating the Q and I signals feeding the color TV set, and the other is for the actual transmission of the Q and I signals as a subcarrier of the slow-scan signal.

There are many writeups on the use of the MC 1596. Basic information is found in the *ARRL Handbook* and also in application sheets for the MC 1596. It is one of these application-note circuits that is used in the color balanced modulator shown in Fig. 6-49.

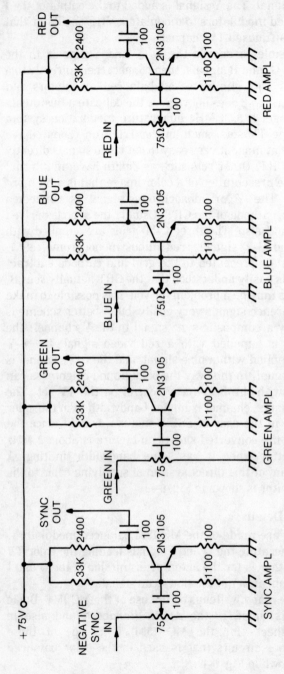

Fig. 6-48. One method of supplying color to a fast-scan monitor (Zenith). The exact connections will vary from model to model.

The use of balanced modulators in SSB and color systems is well established. Basically, two channels can be transmitted in the same frequency band if one of the balanced modulators is fed with a 90° carrier frequency. The other balanced modulator is fed with the same frequency but at 0° phase. SSB enthusiastics should realize if one more 90° phase shift is put into one of the video or audio channels, one of the sidebands can be cancelled. This is known as the phasing system of creating SSB. The Central Electronics 10A, 20A and 100V and the Hallicrafters HT-37 are examples of this method of generating SSB signals.

Balanced modulators are easily tested by putting in a single audio tone into the video or audio input. This creates a DSB signal that can be used for proper signal adjustment. This is very well described in the *ARRL Handbook*.

In all systems that use quadrature modulation, it is necessary to have a means of carrier phase reference. In color TV, this is done by using a color burst during a portion of the horizontal sync time. This is also the system used in the slow-scan color converter, except the burst is kept *high* or *on* during the entire sync time. This is an economical method of minimum circuit complexity. In the case of color transmission, a 500 Hz carrier is used instead of the 3.58 MHz. In fact, some thought was given to using the same modulator that is used to encode the signal being fed to the color TV set, but unnecessary circuit complexity resulted. Again the 500 Hz, 0°

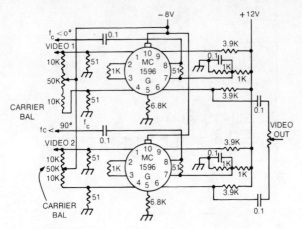

Fig. 6-49. One application of the MC 1596 is in this color balanced modulator.

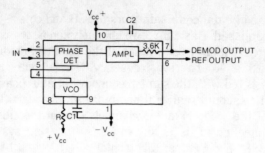

Fig. 6-50. The 565-PLL chip used for slow-scan color synchronization and subcarrier generation.

and 90° phases are generated for modulating the subcarrier. This oscillator is a 500 Hz Signetics 565 IC. It serves two functions; the first is the generation of a 500 Hz signal, and the second is the recovery of the reference signal during reception time by its PLL functions. Any specification sheet on the Signetics 565 will give plenty of details on the use of this circuit. The basic circuit is shown in Fig. 6-50.

Finally, the 3.58 MHz encoded fast-scan picture is radiated into the TV set by a coaxial cable from a 2N918 oscillator, first appearing in a TV typewriter article in *Radio Electronics*, September 1973. This little circuit permits the modulation of the Y channel, I and Q channels, and sync burst on to the chosen VHF channel frequency. The basic circuit is shown in Fig. 6-51.

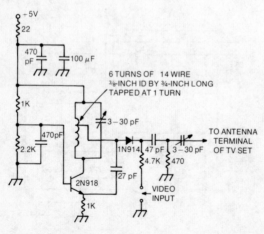

Fig. 6-51. An RF modulator and VHF oscillator for radiating the encoded signal into a fast-scan color TV.

The method of creating the quadrature phases and the 57° phases is unique since the two 0° and 90° phases are generated digitally. The frequencies required to do this bit of phase generation requires two frequencies related by a factor of two. The circuit is shown in Fig. 6-52. The 57° phase shift is created by inserting an RC circuit in the input of one logic gate. Several other methods of creating the 57° were tried by using monostable time-delay circuits. Circuit reliability became a problem, so the RC circuit is used.

The power-supply requirements are not modest. The +5V and −5V supplies need to be at least 5A. The +12V and −12V are 1A supplies. All voltages are developed by the use of 3-terminal T03 regulators made by National of the 309, 320, and 340 types. Quite a lot of heat is developed in the power supply, so it is advisable for the power-supply chassis to be separate from the color converter.

The memory itself consists of 12 boards (WA9UHV source) that contain the 3-color 4-bit plane memories. Each board contains one bit plane that consists of 16,384 bits, input and output interface, and clocking. Two muffin fans circulate air through the memory boards (shown in Fig. 6-53).

The circuits details are complex. Anyone wishing to receive a continuously updated version of the color converter can contact W9NTP to obtain a data package. A nominal charge of $1 will be made for the duplication charges. Write to: Don Miller, P.O. Box 75, Waldron, Indiana 46182.

It is my desire to establish the color slow-scan standards within 1 year of publication of this book. Anyone wishing to suggest standards should contact W9NTP at the above address.

This scan converter was demonstrated at the Dayton Hamvention on April 26, 1975 in Dayton, Ohio. At that time color slow-scan tapes, provided by WB8DQT, were used to load the memory manually, this required a minimum of 24 seconds;

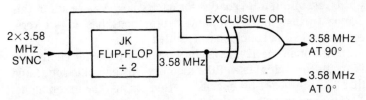

Fig. 6-52. Circuit for digital generation of two quadrature signals for color balanced modulators.

Fig. 6-53. Two views of the slow-scan color converter.

it is possible to load the memory in 8 since the memories are completely parallel. Each of the colors were shown in full 4-bit words, resulting in 16 shades for each color. As a result of this demonstration and comments from thousands of viewers, some interesting conclusions resulted.

First of all the 128 by 128 by 4 memory plane for each color is an uneconomical use of the memory. Instead, it is suggested that 8 of the 16,384-bit boards be used for just the green alone. This means that the horizontal pixel resolution in green is 256 with the usual doubled 128 (pseudo 256) number of lines. This

Fig. 6-54. Dr. Don Miller, W9NTP, at the controls.

will result in excellent frame-grab pictures; and also since most of the information is in green, this will result in a much finer picture. The remaining 4-bit plane boards will be distributed in the red and blue channels. Three of the boards will be used in the red channel that results in a 128 by 128 by 3 picture. Again the number of lines is doubled by presenting each line twice, but the picture has only 128 horizontal pixels and 8 shades in the red. Finally, the remaining bit plane of the original 12 is used as a 128 by 128 by 1 memory for the blue. In fact it is not clear from the tests that blue is needed. The same subjective results can be obtained by just putting in a DC blue as background. During the year after publication it is hoped that enough subjective tests will have been run to positively define all of these standards.

The use of a 500 Hz subcarrier is a sound idea. The information put on these two quadrature channels on the 500 Hz carrier is probably the 3-bit red signal and the 1-bit blue signal since channel-spectrum widths are minimized by some of the thoughts in the previous paragraph.

In conclusion, slow-scan color is with us. It will be simplified within the next year. Many thanks to W0LMD, WB8DQT, and WA9UHV for their help; W0LMD provided the basic design for the original monochrome scan converter, WB8DQT provided some needed color slow-scan tapes that made the Dayton demonstration possible, and WA9UHV laid out the artwork on much of the circuitry that is common between the W0LMD slow- to fast-scan converter and the W9NTP color slow- to fast-scan converter. (See Fig. 6-54.)

Chapter 7
Roundup of Existing Gear

During the early days of SSTV all video gear was do-it-yourself equipment. Amateur radio supported only a minimum number of SSTV operators and growth was somewhat limited. Several years elapsed before commercial manufacturers began to seriously consider entering the SSTV field. As with any area of amateur radio, some of the manufacturers decided to retract from the slow-scan field while other manufacturers became quite popular.

The SSTV world is presently expanding at an exponential rate, which make manufacturing possibilities limitless. This means that several companies may enter the SSTV field in the near future. Notwithstanding future evolvement, this chapter considers the commercial SSTV market as of mid-1975.

There are presently four primary companies producing SSTV gear: Robot Research, Venus Scientific, Sumner Electronics and Engineering, and Sideband Engineering. All of these companies produce top-quality gear that performs exceptionally well for SSTV purposes. In an attempt to deliver the most outstanding unit possible at a reasonable price, all available SSTV monitors utilize solid-state circuitry with a P7 CRT. The same monitor considerations given in Chapter 5 also apply to commercial SSTV units. All available commercial SSTV cameras employ fast-scan sampling techniques. This procedure requires using a vertical fast-scan rate for horizontal slow-scan sweep. Sampling has proven to be a reliable and inexpensive method of SSTV picture generation. Presently, commercial digital scan converters are too expensive for large-scale manufacturing. This situation is due to the large MOS shift-register memory required for direct scan conversion. Possibly scan conversion will become quite popular as shift-register costs decrease.

BUYING CONSIDERATIONS

Many of today's radio amateurs are hybrid builders—they construct some of their gear and purchase some of their gear. The best approach to this situation is to buy a good SSTV camera and build the SSTV monitor. Slow-scan monitors are relatively straightforward in design and easy to align. Contrasted to this, a slow-scan camera utilizes sophisticated circuitry and are very tedious to adjust.

Today's fast-pace living has brought about another type of ham operator—one who would like to build gear but has extreme difficulty merely finding time for on-the-air activities. This individual purchases ready-to-use SSTV rather than completely forgetting SSTV activity. Quite often this individual plays a vital role in the SSTV world—productive use of SSTV as a communications tool. He may work in a field unrelated to electronics, yet his knowledge, shared through the medium of SSTV, can benefit countless others. An example of this situation might be a mechanic showing others how to connect linkages on a carburetor for better gas mileage, or an American farmer describing and sketching advanced irrigation methods for viewing by people in remote areas of the world. The advent of commercial SSTV gear has, indeed, filled a tremendous void in the area of worldwide communications.

Information in this chapter is presented in a systematic method—monitor details, camera details, and accessories. You will investigate the procedures involved in using each of these units and its theory of operation. Block diagrams are included for ease in understanding how each unit operates.

ROBOT SSTV MONITOR

This monitor (Fig. 7-1) appears to be an exceptionally well constructed unit that utilizes high-quality components throughout. One of the outstanding features of this unit is the

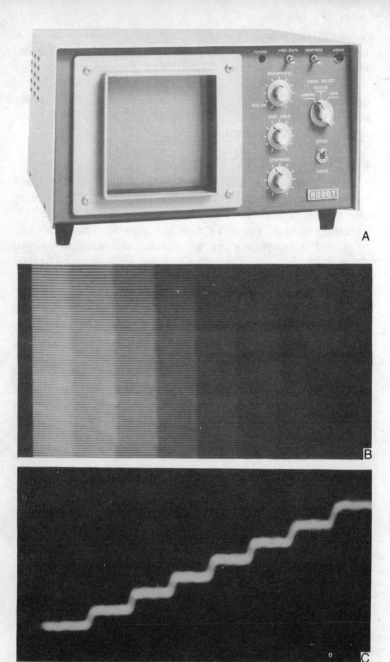

Fig. 7-1. Robot 70 monitor. (A) Three versions have been produced: 70, 70A, and 70B. (B) Gray-scale display on the Robot 70 in eight levels. (C) Same video displayed as above but in the VIDEO-GRAPH mode.

The following foldout section consists of pages 193 through 208.

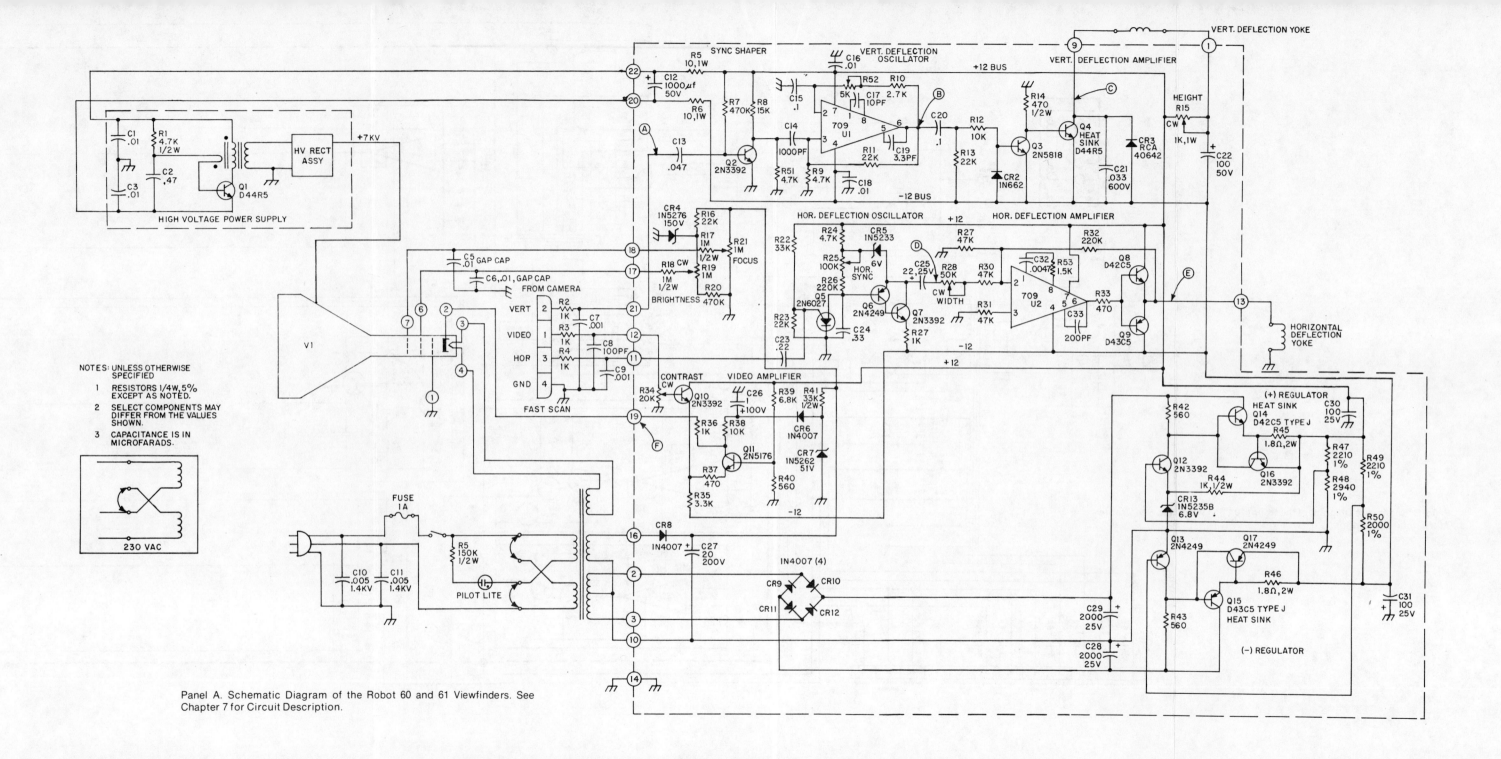

Panel A. Schematic Diagram of the Robot 60 and 61 Viewfinders. See Chapter 7 for Circuit Description.

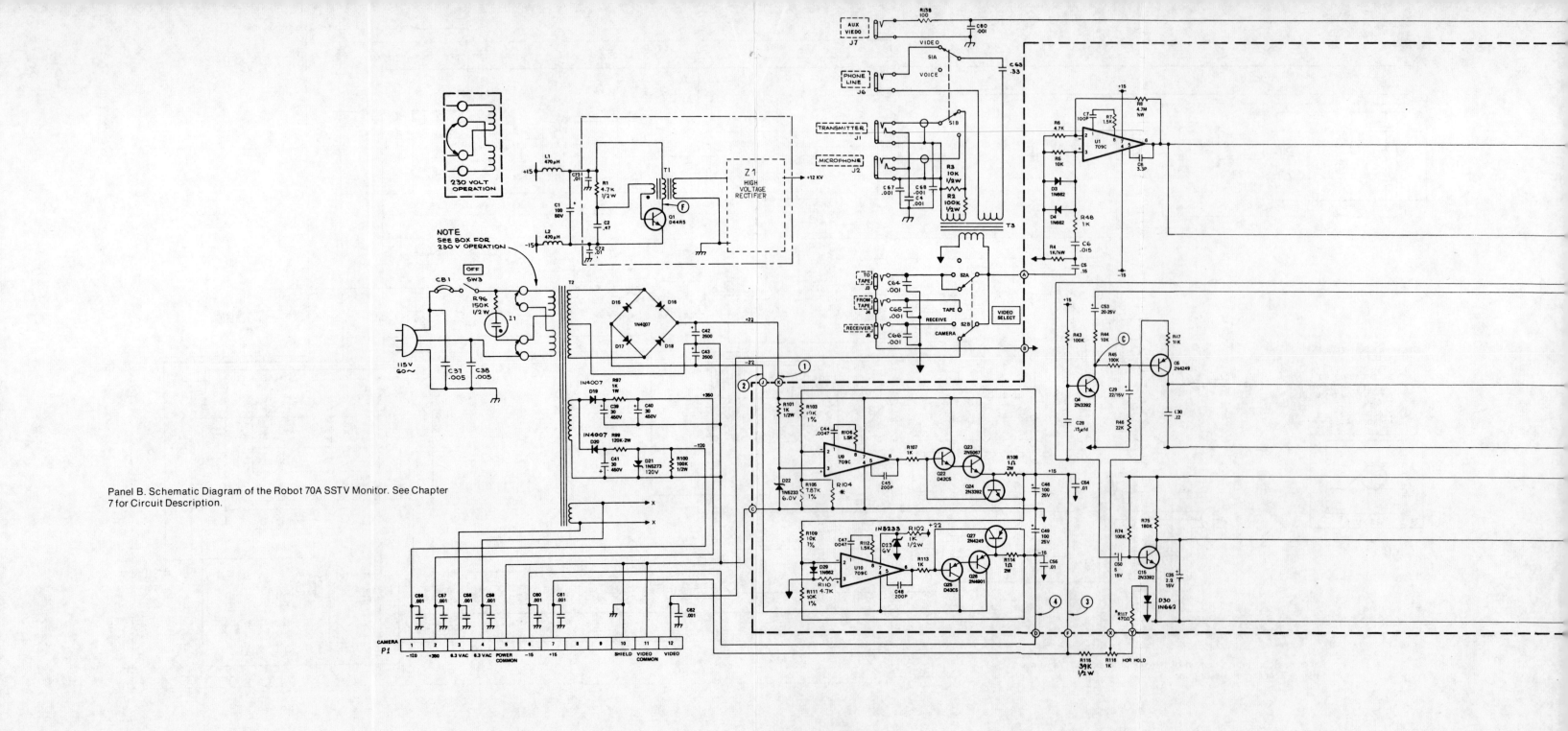

Panel B. Schematic Diagram of the Robot 70A SSTV Monitor. See Chapter 7 for Circuit Description.

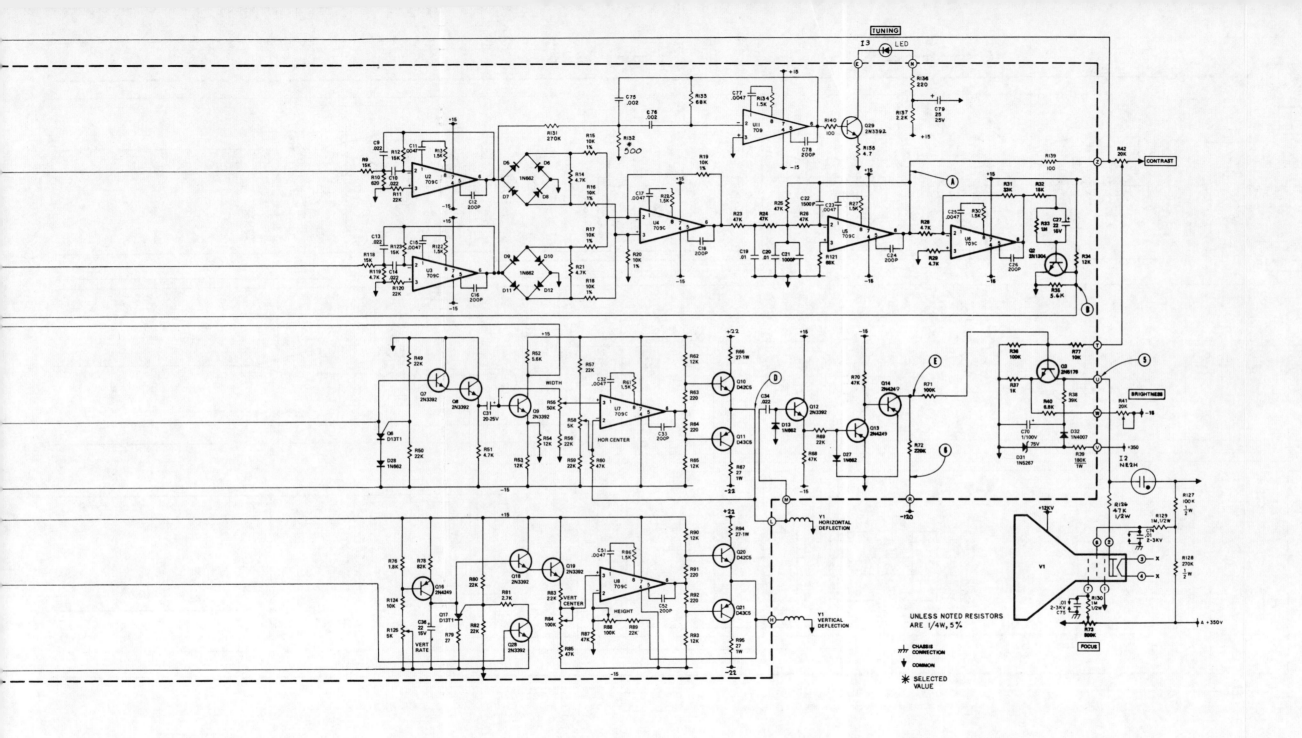

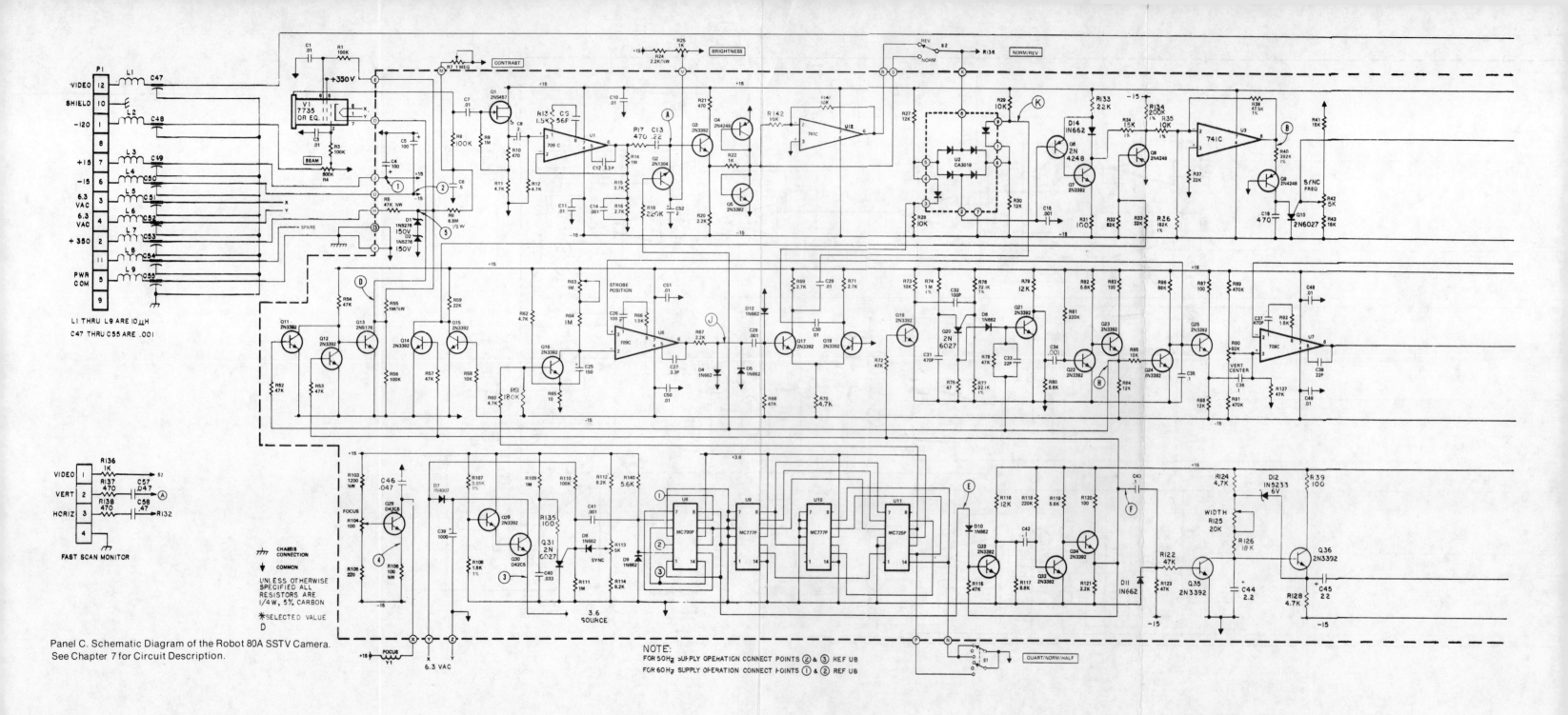

Panel C. Schematic Diagram of the Robot 80A SSTV Camera. See Chapter 7 for Circuit Description.

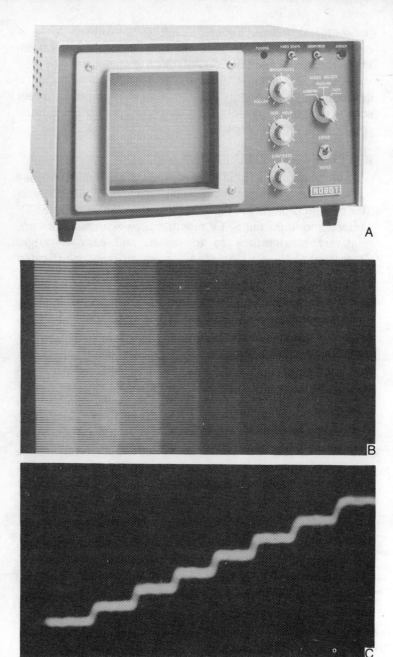

Fig. 7-1. Robot 70 monitor. (A) Three versions have been produced: 70. 70A, and 70B. (B) Gray-scale display on the Robot 70 in eight levels. (C) Same video displayed as above but in the VIDEO-GRAPH mode.

The following foldout section consists of pages 193 through 208.

horizontal slow-scan sweep. Sampling has proven to be a reliable and inexpensive method of SSTV picture generation. Presently, commercial digital scan converters are too expensive for large-scale manufacturing. This situation is due to the large MOS shift-register memory required for direct scan conversion. Possibly scan conversion will become quite popular as shift-register costs decrease.

BUYING CONSIDERATIONS

Many of today's radio amateurs are hybrid builders—they construct some of their gear and purchase some of their gear. The best approach to this situation is to buy a good SSTV camera and build the SSTV monitor. Slow-scan monitors are relatively straightforward in design and easy to align. Contrasted to this, a slow-scan camera utilizes sophisticated circuitry and are very tedious to adjust.

Today's fast-pace living has brought about another type of ham operator—one who would like to build gear but has extreme difficulty merely finding time for on-the-air activities. This individual purchases ready-to-use SSTV rather than completely forgetting SSTV activity. Quite often this individual plays a vital role in the SSTV world—productive use of SSTV as a communications tool. He may work in a field unrelated to electronics, yet his knowledge, shared through the medium of SSTV, can benefit countless others. An example of this situation might be a mechanic showing others how to connect linkages on a carburetor for better gas mileage, or an American farmer describing and sketching advanced irrigation methods for viewing by people in remote areas of the world. The advent of commercial SSTV gear has, indeed, filled a tremendous void in the area of worldwide communications.

Information in this chapter is presented in a systematic method—monitor details, camera details, and accessories. You will investigate the procedures involved in using each of these units and its theory of operation. Block diagrams are included for ease in understanding how each unit operates.

ROBOT SSTV MONITOR

This monitor (Fig. 7-1) appears to be an exceptionally well constructed unit that utilizes high-quality components throughout. One of the outstanding features of this unit is the

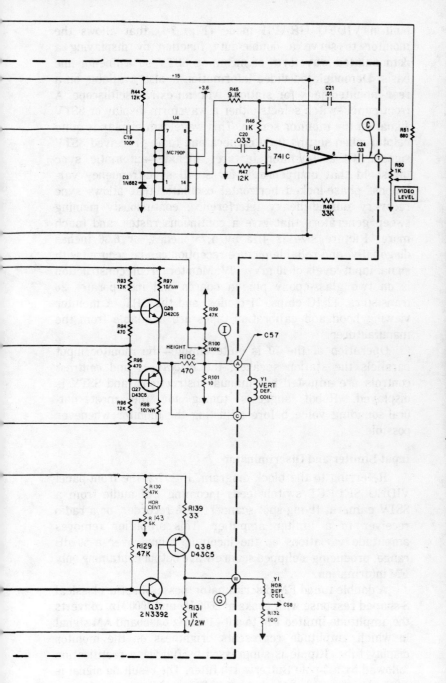

Panel C (cont)

built-in VIDEO GRAPH mode (Fig. 7-1) that allows the monitor to serve a double-duty function by displaying a demodulated waveform of incoming SSTV on a line-for-line basis. Demodulated video information is also available on a rear-mounted jack for stations with an extra oscilliscope. A front-panel switch selects either a waveform display or SSTV signal on the monitor screen. This waveform display is quite helpful when setting up a camera and tuning received SSTV signals. Other monitor features include—automatic sync threshold that compensates for small sync-frequency variations, phase-locked horizontal oscillator that allows sync recovery during heavy interference, continuously running sweep generators that give a continuous raster, and much more. Picture size is 4.75 by 4.75 inches, or 6.36 inches diagonally, and reliable picture reception can be secured with signal input levels of 40 mV−10V. Monitor circuit construction is on two glass-epoxy plug-in boards and incorporates 36 transistors, 12 IC chips, 31 diodes, and the CRT. A monitor viewing hood and calibration tapes are available from the manufacturer.

Operation of the 70 is quite simple—the monitor input parallels the station speaker, the *brightness* and *contrast* controls are adjusted per manual instructions, and SSTV is displayed. Robot suggests tuning for the most natural-sounding voice before adjusting the monitor whenever possible.

Input Limiter and Discriminator

Referring to the block diagram, Fig. 7-2, the front-panel VIDEO SELECT switch feeds incoming FM audio from a SSTV camera, flying-spot scanner, tape recorder, or a radio receiver to a limiter amplifier. This amplifier removes amplitude variations on the incoming signal over a 50 dB range, producing a clipped square-wave output containing only FM information.

A double-tuned FM discriminator, possessing the classical S-shaped response with peaks at 2600 Hz and 1000 Hz, converts the amplitude-limited FM to a 0−1000 Hz baseband AM signal in which amplitude represents brightness on the monitor display tube. Ripple is suppressed by full-wave rectification followed by a 3-pole Butterworth filter. The resulting signal is applied to the CRT cathode via a video amplifier.

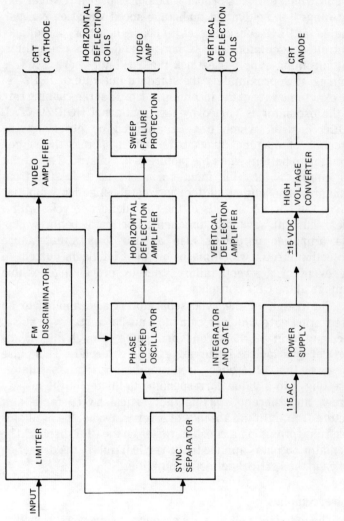

Fig. 7-2. Functional block diagram of the Robot 70.

Sync Separator

The AM video signal is also applied to the sync separator, which removed the blacker-than-black 1200 Hz horizontal and vertical sync pulses inserted into the original picture signal generated in the camera. It employs a self-biased level-gating circuit which extracts proper sync pulses over a receiver tuning range (or equivalent audiotape speed variation ranges). One is a PLL consisting of a phase detector, a voltage-controlled oscillator, and a horizontal-deflection amplifier. The incoming sync pulses lock the oscillator—normally free running at approximately the standard line rate of 15 Hz—to the exact line rate of the incoming video. The free-running rate of the oscillator is varied by the front-panel HORIZONTAL HOLD control, which can be adjusted by observing the direction of diagonal black bars running across the screen from top to bottom when the picture is not in sync. The circuit can maintain *lock* for input variations up to ±20% from standard. The output of the horizontal-deflection amplifier drives the horizontal coils of the CRT deflection yoke with a linear current sawtooth, producing horizontal beam sweep. The amplifier output is also applied to a sweep-failure protection circuit which biases off the CRT beam current in the event of a sweep failure, thereby preventing possible damage at the CRT phosphor.

The output of the sync separator is also applied to an integrating gate which rejects the 5 ms horizontal sync pulses occuring every 1/15 second and gates through the 66 ms vertical sync pulses occurring every 8.5 seconds. This pulse triggers the otherwise free-running vertical oscillator, resetting it to a value corresponding to the top of the display screen and thereby starting the vertical sweep for a new picture frame. In the absence of a vertical sync, the oscillator free runs, providing a constant raster on the CRT display. The oscillator output is applied to the vertical coils of the deflection yoke via the vertical-deflection amplifier.

Power Supplies

The following voltages are produced by the power supply:

- +22V DC
- −22V DC
- +15V DC regulated
- −15V DC regulated
- +350V DC
- −120V DC regulated
- 6.3V AC

The principal power-supply voltage employed by the circuits above is the +15V and −15V. This supply is regulated to better than 0.1%, load and line, and is protected against over-current surges exceeding 0.6A by a protection circuit which reduces the output voltage in the presence of overloads.

High voltage for the CRT anode (12 kV) is provided by a DC-to-DC converter operating from the regulated +15V and −15V supply so that the brightness of the display is free of line-voltage fluctuations. The converter, consisting of a 25 kHz oscillator and a peak voltage doubler, is completely shielded in a separate internal enclosure to eliminate possible interference.

All active components in the 70 are mounted on a glass-epoxy PC board located just under the removable top cover for easy servicing and adjustment. In addition to the front- and back-panel controls, adjustments located on the circuit board include: picture *height* and *width*; picture *centering* (horizontal and vertical); free-running vertical oscillator rate.

Tuning Characteristics

It is helpful to understand the effect of tuning changes on monitor performance. Recall that video signal consists of a series of audio tones ranging from 1200 Hz for sync to 2300 Hz for white. The monitor converts these tones to a voltage proportional to frequency in a discriminator circuit. Voltage-frequency characteristic of the discriminator is S-shaped in such a way that signals outside the normal range are interpreted as gray, rather than either black or white. With correct tuning, the sync frequency (1200 Hz) is located near the upper peak of the curve and the white frequency (2300 Hz) is located near the lower peak. All frequencies between these two lie on the straight portion of the curve. (See Fig. 7-3).

If tuning is too low, the sync frequency will produce a lower voltage than black frequency (1500 Hz). Because the sync separator operates by picking off the most positive voltage, this degree of mistuning will cause the monitor to sync on black portions of the picture. The effect of this amount of mistuning will be observed in the picture as a loss of sync. When this occurs, tune to produce a higher pitched sound.

Tuning too high in pitch will cause sync (1500 Hz) to produce the most positive voltage, thus the picture will sync

Fig. 7-3. A typical 128-line display with an 8-second time exposure from the Robot 70 monitor screen.

properly; however, white (2300 Hz) will be pushed over the peak causing white tones to be reduced in voltage. The effect of this tuning is to reduce the observable detail in the white portion of the picture.

When strong adjacent-channel interference is present it will sometimes be helpful to turn off the AGC of the radio receiver. The monitor contains a limiter which removes the effect of large changes in signal amplitude.

Operation With the Telephone Line

To talk place the VIDEO/VOICE switch in the VOICE position. SSTV pictures from the local camera or tape recorder may be viewed on the monitor during a conversation. Place the VIDEO SELECT switch in the TAPE position to view the local tape recorder or in the CAMERA position to view the local camera.

To receive pictures place the VIDEO SELECT switch in the RECEIVE position and the VIDEO/VOICE switch in the VIDEO position. Adjust BRIGHTNESS and CONTRAST to suit local lighting conditions.

To send pictures place the VIDEO/VOICE switch in the VIDEO position. Place the VIDEO SELECT switch in CAMERA position to send pictures from the camera. Place the VIDEO SELECT switch in the TAPE position to send pictures recorded on the tape recorder.

When sending pictures from the tape recorder it is necessary for the tape recorder to produce about 2V P-P across the telephone line. The 70 places a 1-to-1 transformer between the tape recorder output and the phone line. Depending on the choice of tape recorder, it may be necessary to install either a matching transformer or amplifier between the tape recorder output and the monitor.

Caution—it is against the law to record telephone conversations without the express permission of the person being recorded.

ROBOT 80A SSTV CAMERA

The 80A camera (Fig. 7-4) utilizes a popular 7735A vidicon to produce high-quality TV pictures. This tube is operated at a fast-scan rate and the sampling principle (as described in Chapter 2) is employed to produce SSTV. An internal modulation calibrator is incorporated for establishing black and white levels; and frame lengths of 1/4, 1/2 or full raster are selectable via a rear-panel switch. A NORM/REV video switch and a viewfinder fast-scan output jack are provided. SSTV output level is controlled via a rear-panel potentimeter. Power for the 80A camera is supplied by the Robot monitor, and all associated cables are included with the camera. Circuit construction is on a single glass-epoxy plug-in board. The unit incorporates 40 transistors, 12 IC chips, 7 diodes, and the vidicon. Robot has a selection of eight various lens available for use with the camera, or you may use your own lens on the standard C-mount.

Operation of the 80A camera is relatively straightforward after the function of each control is understood. Camera

Fig. 7-4. Front view of the Robot 80 camera. A NORM/REV switch to provide white-on-black lettering.

BRIGHTNESS and CONTRAST controls are adjusted per manual instructions, then *beam current* and *video levels* are set. Next, the camera lens is adjusted for proper focus and illumination.

Video Chain

Referring to the block diagram, Fig. 7-5, fast-scan video (4 kHz line and 15 Hz frame) is generated by the vidicon camera tube. A video amplifier consisting of a FET input stage and an

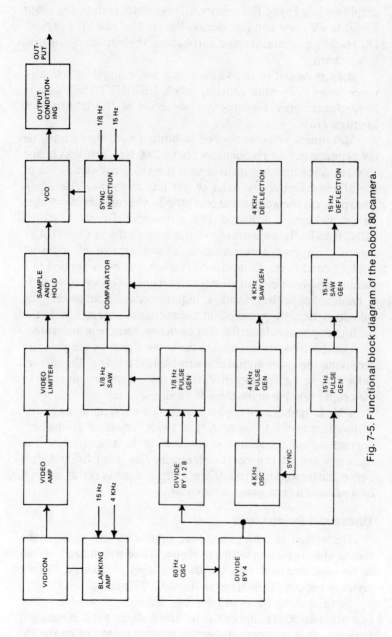

Fig. 7-5. Functional block diagram of the Robot 80 camera.

operational amplifier increases the signal level to approximately 1V over a 250 kHz band. Following the video amplifier is a keyed DC restorer. This circuit clamps the video signal to 0V once per line during the retrace blanking period. DC coupling is maintained throughout the remainder of the video chain.

Bias is added to the video signal for manual insertion of black level. The bias control, labeled BRIGHTNESS on the front panel, provides operator selection of the transmitted average gray.

Maximum video excursion is limited such that whites do not produce output frequencies above 2300 Hz nor blacks below 1500 Hz. This limiting insures that the video will not go out of the allowed frequency band or get into the sync. Video-gain control is via the adjustable lens stop or vidicon target voltage. Target voltage is adjusted with a front-panel control labeled CONTRAST. By adjustment of the lens stop and CONTRAST control you can accommodate a wide range of external lighting conditions. Conditioned fast-scan video is applied to a sample-and-hold circuit which consists of a 6-diode IC switch, a holding capacitor, and a high-impedance follower. The 6-diode switch offers excellent bidirectional charge flow to the holding capacitor, insuring that each new sample is accurate.

Composite sync is added to the video signal by temporarily removing the video signal and substituting a bias of the correct value. A fixed-gain amplifier increases the composite video to the proper level to drive the FM oscillator.

FM is generated by a controlled-current source and unijunction oscillator operating at twice the output frequency. Correct output frequency is obtained by binary division. The square wave at the correct frequency is band limited in a 2-pole Butterworth active filter. Output level is set as desired by means of a rear-panel attenuator.

Timing and Sweep Chain

Deflection, blanking, and sync signals are generated in the timing chain starting with a 60 Hz oscillator which can free run or be synchronized with the 60 Hz supply frequency. Binary dividers reduce the 60 Hz to 15 Hz and 1/8 Hz rates.

Note that 15 Hz is obtained by dividing 60 Hz by 4. For operation on 50 Hz supply mains, the division ratio is changed to three. This is accomplished by moving a jumper on the PC board.

For mobile operation, where the supply frequency is not stable, the internal 60 Hz oscillator can be used as a reference. This is accomplished by removing a resistor from the PC board.

From the 15 Hz timing signal, a one-shot multivibrator forms a 5 ms pulse. This pulse performs the following functions:

- Drives the 15 Hz sawtooth generator
- Blanks the vidicon frame retrace
- Contributes to the transmitted sync
- Synchronizes the 4 kHz sweep generator

A switch-capacitor sawtooth generator provides 15 Hz deflection, driving an operational amplifier and follower-feedback pair. The 4 kHz deflection chain consists of a unijunction oscillator synchronized by the 15 Hz pulse. Synchronization between the 15 Hz and 4000 Hz is used to prevent a random relation between frame and line scan at the beginning of each frame. A trigger pulse, taken from the oscillator, drives a single-shot multivibrator to produce a 4 kHz pulse which blanks the vidicon and drives the 4 kHz sawtooth generator. The 4 kHz sawtooth and deflection circuits are similar to those used in the 15 Hz deflection.

A 1/8 Hz pulse, 66 ms wide, is formed in a logic gate which derives its inputs from the divider chain. The 1/8 Hz pulse drives the 1/8 Hz sawtooth generator and contributes to the transmitted sync. The sawtooth generator is of the switch-capacitor type.

The sample-and-hold switch is operated by a narrow pulse generated when the amplitude of the 1/8 Hz and the 4 kHz work a lens with a sensitivity of at least $f/1.9$ is desirable. For outdoor daytime use a maximum sensitivity of $f/3.5$ is adequate.

Lens Characteristics

The lens focal length is determined by the size of the televised image. A given subject at a given distance determines the focal length to be used. A simple relationship between these quantities is given by the following equation:

$$fl = 10d/s$$

Where:

fl = focal length in millimeters
d = camera to subject distance
s = subject size in same units as d

For example, suppose it is desired to televise a face which is about 1 foot high and the desired distance is 2.5 feet. From the above formula the focal length is determined to be 25 mm. Lenses are available in a wide range of focal lengths to suit problems from close up to distance. The 80 accepts the industry-wide C-mount (32 threads per inch).

Normally the distance between the camera tube and the lens is adjusted so that the focus marks on the lens are correct; however, if it is required to use a lens at an unconventional distance it is possible to use a different spacing. For example, a 25 mm lens is usually marked for focus down to 1.5 feet; however, by increasing the spacing between the lens and the camera tube, working distances as short as a few inches are possible. Again, the formula may be used to determine the required quantities. For example, it is desired to televise a 2.5-inch picture with a 25 mm lens such that the picture fills the screen. The formula, arranged to find the distance between the camera and picture is as follows:

$$d = fl\,s/10$$

In the example, $fl = 25$, s = 2.5 inches, and d is therefore 6.25 inches. Since the lens is only marked down to 1.5 feet it will be necessary to increase the distance between the lens and camera tube.

The term C-mount describes the method of attaching the lens to the camera. It also implies that the dimensions of the image formed is suitable for 16 mm movie film. The C-mount lens has long been standard for TV cameras which use the 1-inch diameter vidicon. Its acceptance is due to the fact that the image formed by the lens is just the fight size for the vidicon. Note that the image on the face of the vidicon is 3/8 by 3/8 inch in the 80 camera.

All C-mount lenses are designed such that the distance between the lens mounting surface and the image is a fixed number. For example, this back working distance is the same for a 25 mm and a 200 mm focal-length lens. Because of this design feature it is possible to exchange lens without the need to make any camera adjustments.

C-mount lenses are available in a wide range of focal lengths. Starting at a focal length of 12.5 mm there are lenses which take in a wide-angle field of view. Such a lens is suitable for closeup shots of faces and apparatus. A 25 mm focal-length lens is a good general-purpose choice. It is useful for both closeup and short-distance work. At focal lengths of 50–200 mm are the narrow-angle lens, which are most used for distant scenes. Beyond 200 mm the name telephoto is usually applied, and such a lens is used for extreme-distance work.

Lenses of all focal lengths have calibrated focus from infinite distance to some close distance. The closest distance of focus is typically 1.5 feet for a 25 mm lens and greater for longer focal lengths. Thus, the conventional lens is not suitable for closeup work on subject matters such as photographs. Closeup work is accomplished by using any one of the following alternatives:

- A threaded extension barrel between the lens and camera to increase spacing
- Auxillary closeup lens which attaches to the front of the conventional lens
- A macrofocus lens—one which will focus down to 6 inches instead of 18 inches
- Modify the camera to permit the vidicon to be moved farther from the lens

The most convenient of all above alternatives is the macrolens because it provides focus distance calibration marks. With each of the other choices, focus must be accomplished by observing the pictures.

Zoom Lens

The zoom lens provides adjustable focal lengths. For example, one such lens adjusts from 22 to 66 mm focal length. As the focal length varies the lens is kept in focus by internal mechanisms. It is important to note that the focus and zoom track only when the vidicon is placed at the standard working distance. Most zoom lenses are arranged to focus as close as 5 feet. Thus, the lens is not suitable directly for closeup work. The closeup alternatives listed above can also be used with the zoom lens; however, when the zoom lens is used for closeup work the zoom and focus will not track.

Fig. 7-6. Front view of the Robot 61 viewfinder used for display of video taken by the Robot 80 or 80A cameras.

ROBOT 60 AND 61 VIEWFINDERS

These units display fast-scan pictures generated by the Robot 80 or 80A camera. This allows one to easily and quickly set camera adjustments, establish proper gray shades, and view picture content. As the units are RFI proofed, they can be used during SSTV transmissions. The viewfinders operate at the fast-scanning rates of the camera vidicon, which means 15 frames per second are displayed. As you know, conventional fast-scan TV operates at 30 frames per second to eliminate flicker. The viewfinders overcome this problem by using P7 CRT for display. Thus, only a slight flicker is noticeable during operation. The 60 viewfinder can be made to fit on the rear of a Robot camera if desired. Picture size of this unit is 4 inches diagonal. The 61 (Fig. 7-6) is somewhat larger, allowing ideal desk-top operation. Picture size of the 61 is 6.5 inches diagonal. Both units are furnished with cables for direct plug in operation with Robot cameras.

Operation of either viewfinder is the essence of simplicity—just plug it into the AC line and camera, then switch it on. All controls of the viewfinder are internal; if the controls were available to the operator its function as a standard would be destroyed.

Vertical Sweep

The sawtooth representing vertical deflection (Fig. 7-7) in the camera is carried to the viewfinder by the connecting cable. A sync separator operates on this waveform to provide an output only during the retrace time of the sawtooth. This pulse output is used to synchronize the free-running vertical oscillator.

The vertical oscillator is a multivibrator constructed with an operational amplifier. The HOR HOLD control sets the frequency of this free-running oscillator. A driver transistor, placed between the multivibrator and output stage, provides the heavy base current required by the output stage. Refer to Fig. 7-8 for a simplified schematic of the vertical output stage.

During period (1) the transistor is like a closed switch and current builds up in the yoke due to the power supply voltage. At the end of period (1) the transistor switch opens and the resonant circuit composed of the yoke and capacitor completes a half-cycle of oscillation. This half-cycle is period (2). Note that during this time the yoke current retraces. At the beginning of period (3) note that the yoke current is at its negative extreme and the yoke voltage has just swung negative. The diode now conducts and the yoke current dies toward zero exponentially. Just before the yoke current reaches zero the transistor switch closes and current is forced into yoke by the power supply.

Horizontal Sweep

The sawtooth representing horizontal deflection in the camera is carried to the viewfinder by the connecting cable. The sawtooth is applied directly to the free-running horizontal oscillator. During retrace of the sawtooth the horizontal oscillator is also forced to retrace.

The horizontal oscillator is bootstrapped to provide a linear sawtooth. A linear amplifier, consisting of an operational amplifier and power amplifier with overall feedback amplifies the sawtooth so it will drive the deflection yoke.

Video Amplifier

The 1V P-P video signal is carried from the camera of the viewfinder by the shielded cable. This signal is inverted,

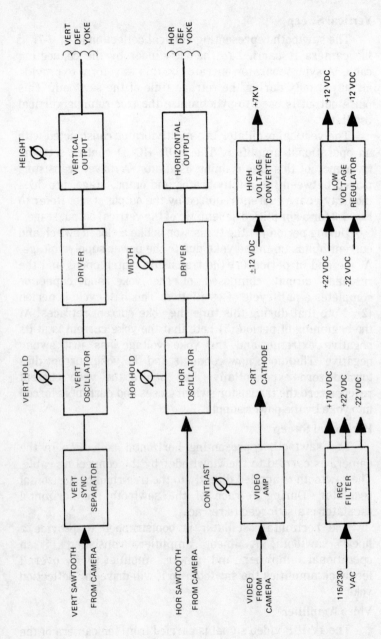

Fig. 7-7. Block diagram of the Robot 60 and 61 viewfinders.

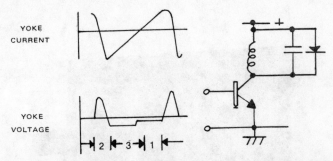

Fig. 7-8. Simplified schematic diagram of the vertical output stages in the Robot 60 and 61 viewfinders.

amplified, and applied to the cathode of the CRT. An internal contrast control varies the gain of this amplifier.

Power Supplies

Power-supply voltages are +12V DC, −12V DC, +7 kV, 170V DC and 6.3V AC. These voltages are derived from the AC power line. The +12V DC and −12V DC are regulated. The +7 kV is produced by a DC-to-DC converter supplied from the regulated 12V DC.

ROBOT 300 SSTV SCAN CONVERTER

This recent Robot item (Fig. 7-9) utilizes an analog storage tube to permit both fast- to slow- and slow-to-fast scan conversion. This method allows standard fast-scan gear to be used directly on SSTV. A conventional camera drives the unit during transmissions, then during reception SSTV pictures may be viewed on an unmodified home TV. The unit includes frame grab on both transmit and receive, video inversion, partial frame scan, and practically anything else the avid SSTV operator can imagine. There is also 256-line capabilities, so that stations equipped with the 300 converter can exchange high-resolution pictures. (See Fig. 7-10.)

ROBOT 400 CONVERTER

Robot's latest addition to their superb line of SSTV equipment is the fully solid-state 400 scan converter (Fig. 7-11). This unit includes controls and indicators for setting the stored gray level range of the *snatched* TV picture, and for setting the stored gray level range of the received SSTV picture. A built-in gray-scale generator provides a ready standard for accurate setting of transmit and receive controls.

Fig. 7-9. Robot 300 scan converter. This unit permits fast-to-slow scan conversion as well as slow-to-fast scan conversion.

Fig. 7-10. Monitor displays of scan conversion performed by the Robot 300. (A) 128-line picture with average resolution. (B) The same shot with a 256-line display with better resolution.

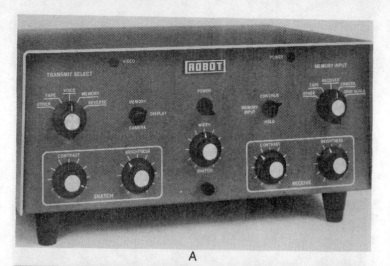

Fig. 7-11. Robot 400 scan converter. (A) Front panel of the Robot 400 scan converter. (B) Memory-stored display of a snatched picture taken by the Robot 400.

Installation for operating with a radio station consists of connecting a single cord between the speaker and output of the radio receiver and the FROM RADIO jack on the Robot 400.

To transmit SSTV pictures, connect a shielded 2-wire cord between the TRANSMITTER jack on the Robot 400 and the microphone input to the radio transmitter. The station microphone is plugged into the MICROPHONE jack; selection between SSTV and voice is by means of a front-panel switch.

The heart of ROBOT 400 is a 65,536-bit solid-state memory made up of 16 random access memory (RAM) 4096-bit chips. The displayed picture consists of an array of 128 by 128 discrete picture elements, each element coded into one of 16 gray shades. Because of the discrete array of picture elements and discrete gray shades, digital quantization effects are visible upon close examination of the picture. A typical picture stored in the memory of the Robot 400 and displayed on a standard TV set is reproduced in Fig. 7-11B.

VENUS SS2 MONITOR

The SS2 monitor (Fig. 7-12) is a compact and high-quality unit utilizing several unique features. A front-panel switch selects either SSTV or spectrum-analysis display (ACCU SYNC) on the monitor screen. The spectrum analysis is on a line-for-line basis (Fig. 7-13). This analysis display is also quite advantageous for tuning received SSTV stations. LED (light emitting diode) indicators are placed at strategic circuit points for aid in instantly determining any possible failure.

Fig. 7-12. Front view of the Venus SS2 SSTV monitor. This monitor features an Accu-Sync mode to display sync pulses on the CRT.

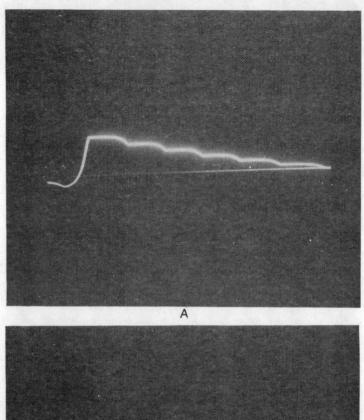

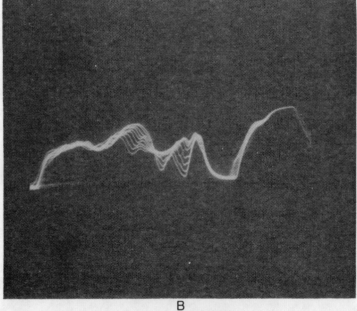

Fig. 7-13. Accu-Sync displays on the SS2 monitor. (A) This display is during a gray-scale pattern received by the monitor. (B) This Accu-Sync display is while receiving actual video on the SS2.

Each monitor is supplied with a personalized name tag that mounts on the front panel and a collapsible foot provides comfortable viewing angles.

The SS2 utilizes a combination of free-running sweep and time-triggered sync to provide reliable operation during adverse propagation times. Automatic CRT burn protection is included in the vent of sweep failure. Picture size is 3¼ by 3¼ inches, or 4.59 inches diagonal. Since the SSTV monitor is usually placed near the viewer, a large screen is unnecessary.

Monitor operation is quite simple: Monitor input parallels the station receiver, then BRIGHTNESS and CONTRAST are adjusted per instruction manual. The monitor operates beautifully with input levels of 40 mV−10V. Either kits or factory-assembled units are available, and each assembled unit is *cooked in* at the factory. Figure 7-14 shows the main PC board of the SS2.

General Description

The SS2 may be operated in either of two modes: SSTV (OPERATE position of mode switch) or ACCU SYNC (ACCU

Fig. 7-14. Main PC board in the SS2. The use of IC chips greatly reduces the size of a monitor.

SYNC position). The following description is keyed to the simplified block diagram of Fig. 7-15 which indicates the basic circuits and the switching arrangement.

Operate Mode

The selected FM video signal is clipped to minimize AM variations and noise, then is applied to two channels: sync and FM demodulation. In the sync channel, the 1200 Hz sync signals are filtered from the composite video/sync signal, detected into pulse form, and amplified. Horizontal and vertical sync pulses are separated and applied to associated sweep generators, synchronizing the horizontal sweep (line rate) to 15 Hz nominal and the vertical sweep (frame rate) to 8.5 seconds nominal.

The sawtooth sweep voltages are amplified and drive the horizontal and vertical deflection coils in the CRT yoke, producing the required SSTV raster. A demodulator converts the FM video into AM video. The amplified video signal is applied to the control grid of the CRT, modulating the CRT beam to reproduce the SSTV image.

Accu Sync Mode

In this unique mode, the SS2 circuits are switched from SSTV to oscilloscope form. At this time, both the video and sync signals are displayed to simplify receiver tuning and troubleshooting. As shown in Fig. 7-15, an inverted version of the horizontal sync pulse drives the horizontal sweep generator. Since the sweep generator is triggered by negative-going transitions of the sync pulse, horizontal sweep is delayed in time by the width of the sync pulse. This arrangement permits the sync pulse to be displayed on the CRT rather than having it occur during horizontal retrace.

With the mode switch set to ACCU SYNC, the vertical sweep amplifier is driven by the demodulated video signal rather than the vertical sweep generator in standard oscilloscope form. In addition, the CRT control grid is grounded to eliminate Z-axis modulation. In this manner, ACCU SYNC operation causes the SS2 to display the video and sync signals at the normal line rate. Flipping the mode switch to OPERATE immediately restores the SSTV display.

Detailed Description—Operate Mode

With the following description of specific circuit areas it is necessary for you to follow the text with the appropriate

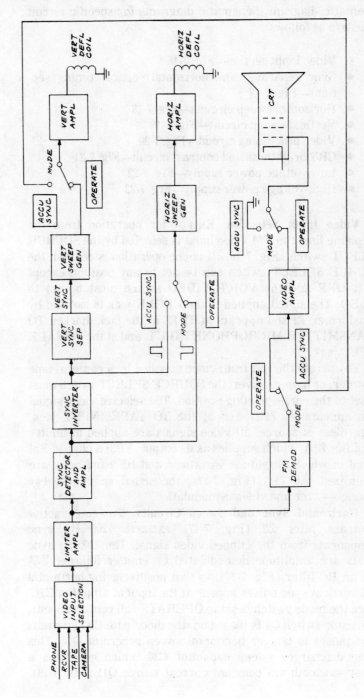

Fig. 7-15. Simplified block diagram of the SS2

233

schematic diagram. Schematic diagrams for specific circuit areas are as follows:

- Video input section—Fig. 7-16
- Sync separator and horizontal/vertical timing sections—Fig. 7-17
- Horizontal sweep circuits—Fig. 7-18
- Vertical sweep circuit—Fig. 7-19
- Video processing circuit—Fig. 7-20
- CRT brightness and contrast circuit—Fig. 7-21
- Low-voltage power supply—Fig. 7-22
- High-voltage power supply—Fig. 7-23

Video Input Selection. Except for operation from the telephone line, the FM video input is selected by the SOURCE SELECT switch (Fig. 7-16). If phone operation is desired, the SOURCE SELECT switch can be set to any position except PWR OFF and the VOICE/VIDEO switch must be set to VIDEO. The signal applied to the PHONE jack is isolated by transformer T2 and appears at TP17, at the jack marked TO TRANSMITTER MICROPHONE INPUT, and at the TO TAPE INPUT jack.

To operate the SS2 from video supplied by a camera, tape recorder, or ham receiver, the SOURCE SELECT switch must be set to the corresponding position. The selected video signal then appears at TP17 and at the TO TAPE INPUT jack. Regardless of source, all video signals are applied to limiter amplifier Z1A, which supplies as its output (TP3) a clipped FM signal in which amplitude variations and HF band noise are minimized. At TP3 (Fig. 7-17), the signal splits into two channels—sync and video demodulation.

Horizontal Sync and Sweep Circuits. Two-stage active bandpass filter Z2 (Fig. 7-17) extracts 1200 Hz sync components from the clipped video signal. The 1200 Hz sync bursts are amplitude demodulated by emitter follower Z7A and an RC filter (Fig. 7-17), so that positive-going horizontal and vertical sync pulses appear at the input of amplifier Z7C. Since the mode switch is set to OPERATE (all contacts open), transistor switch Q4 is biased on, the diode gate CR25 permits sync pulses to trigger horizontal sweep generator Q10. This stage discharges sweep capacitor C36, which generates a linear sawtooth via constant current source Q11 (Fig. 7-18).

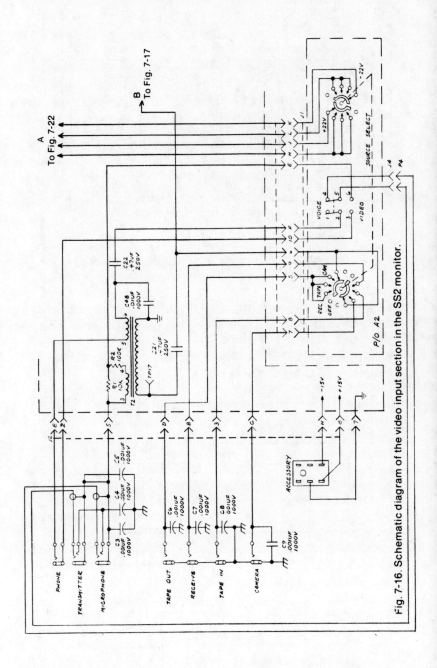

Fig. 7-16. Schematic diagram of the video input section in the SS2 monitor.

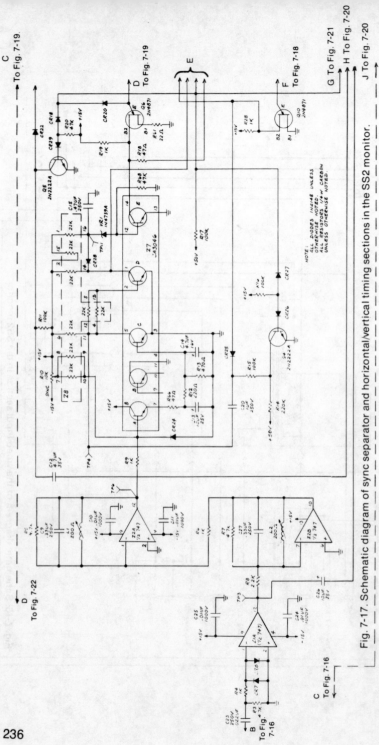

Fig. 7-17. Schematic diagram of sync separator and horizontal/vertical timing sections in the SS2 monitor.

The horizontal sawtooth is applied to sweep amplifier Z3B then through buffers A12 and Q13 to the horizontal deflection coil. Gain pot R50 adjusts horizontal size; offset pot R52 controls horizontal centering.

Vertical Sync and Sweep Circuits. Vertical sync pulses are detected and amplified by Z7A and Z7C (Fig. 7-17), respectively, as described for horizontal sync. Amplifier Z7D inverts the sync signals at TP4; the positive-going sync pulses back bias diode CR28. Receipt of a 30 ms vertical sync pulse cuts off CR28 long enough to permit C15 to charge to the breakdown voltage of VR1. Amplifier Z7E then conducts, triggering Q6, which discharges C16 and generates the vertical sweep. The vertical sweep generator and amplifier circuits are identical to the corresponding horizontal sweep circuits previously described. See Fig. 7-19.

The momentary RE-SCAN position of the mode switch allows the operator to reset the vertical sweep. When the switch is pressed, R18 immediately discharges capacitor C16. When the switch is released, C16 starts to charge, initiating a new vertical sweep.

Video Processing Circuits. The video processing circuits demodulate the FM video signals from the video input selection circuits and amplify the resulting AM video signals to the level required by the CRT control grid for SSTV operation. The clipped FM signals from clipper amplifier Z1A (Fig. 7-17) are fed to a frequency-to-voltage converter consisting of Z6 and the sample-and-hold circuit (Fig. 7-20). Gating circuits in Z6 deliver two pulse outputs. A chain of narrow pulses, at twice the incoming frequency, appears at TP5. An identical pulse chain, slightly delayed in time, appears at Z6-11. The two pulse chains are applied to a sample-and-hold circuit. Each pulse from Z6-11 discharged C29 through Q1. The height of the voltage ramp developed across C29 just prior to discharge represents the video signal amplitude. Switch Q3 passes each peak value of C29 to amplifier Z1B and holding capacitor C28 so that the voltage developed across C28 is a demodulated AM video signal.

The video signal is clamped to ground by C33 and CR10, a feature that permits independent brightness and contrast control. After amplification by Q15 and Q16, the video is applied to the CONTRAST control, which sets the video level applied to the CRT grid.

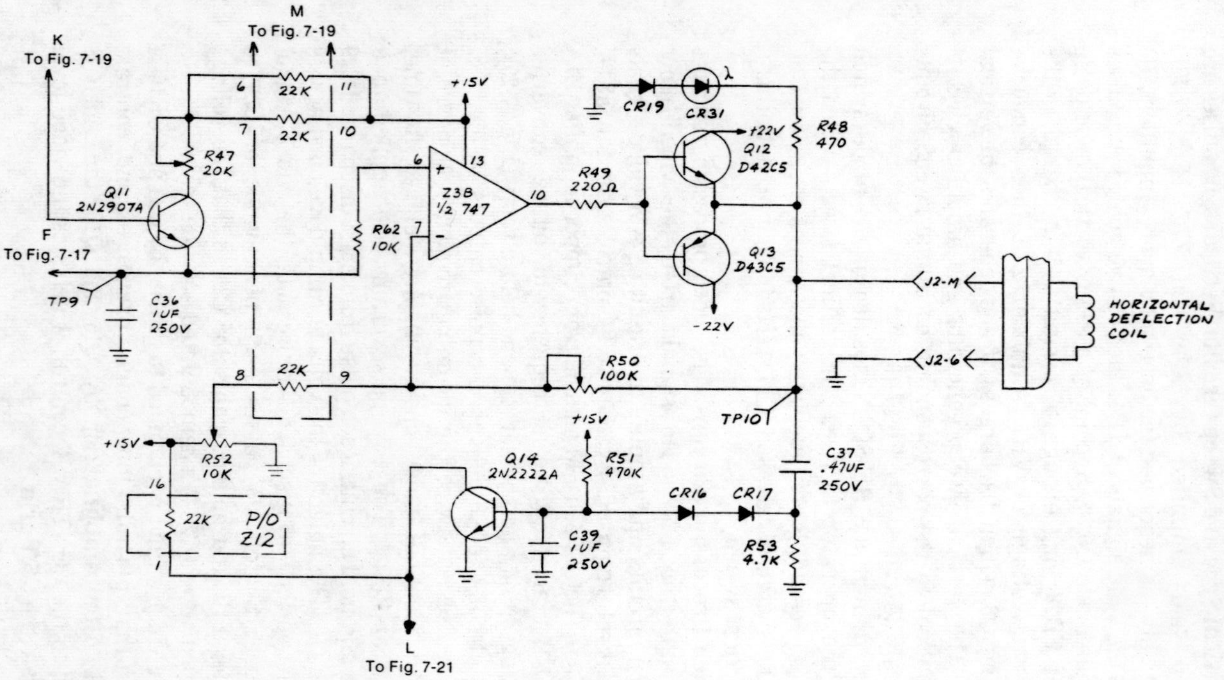

Fig. 7-18. Schematic diagram of horizontal sweep circuit in the SS2 monitor.

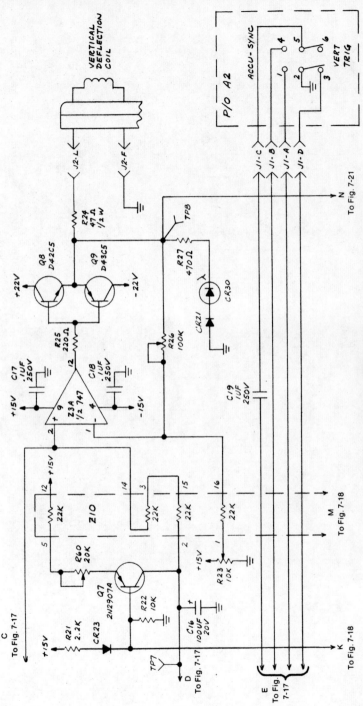

Fig. 7-19. Schematic diagram of vertical sweep circuit in the SS2 monitor.

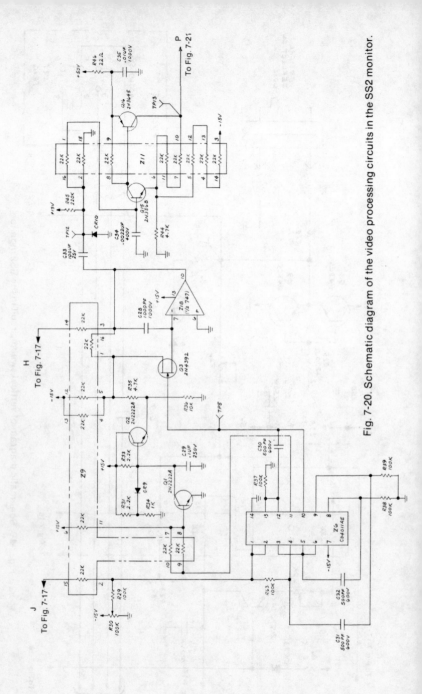

Fig. 7-20. Schematic diagram of the video processing circuits in the SS2 monitor.

CRT Brightness Control Circuits. The BRIGHTNESS potentiometer provides a variable source of DC voltage for inverter Q17 (Fig. 7-21) which in turn drives the CRT cathode through current limiting circuit R58, Q20, and Q21. CRT cathode voltage is adjusted by the BRIGHTNESS control. Under normal operating conditions, the current limiter acts as a short circuit. When current increases above 20 μA (nominal), the current limiter becomes active and automatically limits the current, protecting the CRT. Vertical retrace blanking is accomplished by turning off Q17 via CR15 at retrace time.

Sweep monitor Q14 (Fig. 7-18) protects the CRT in the event of horizontal sweep failure. If the horizontal sweep should fail, Q14 conducts, and DC voltage is removed from the BRIGHTNESS control. The CRT cathode voltage then rises, turning off the display and protecting the CRT from burnout.

Low Voltage Power Supply. A full-wave rectifier, consisting of diodes CR1 through CR4 and filter capacitors C1, C2, C7, and C8 provide +22V DC and −22V DC (Fig. 7-22). These voltages are present as long as the AC line cord is plugged into a source of AC power. Filament voltage (10V nominal) for the CRT is also applied to the CRT at this time, permitting instant-on operation when power is applied. Setting the SOURCE SELECT switch away from the PWR OFF position causes +22V and −22V power to be fed to regulators Z4 and Z5, which provide +15V and −15V to the SS2 circuits.

High Voltage Power Supply. High voltage is generated by 20 kHz oscillator Q18 and Q19 operating in conjunction with a high-voltage transformer (Fig. 7-23). The transformer drives a voltage multiplier that delivers 6V for the second anode of the CRT. The voltage developed across other transformer windings are rectified into +400V and +50V.

Detailed Description—Accu Sync Mode

Since most of the circuits in the SS2 are unaffected when switching the mode switch from OPERATE to ANNU SYNC, the following description covers only the circuit changes introduced by this tuning and diagnostic mode.

Vertical Sweep and Video Circuits. When the mode switch is set to ACCU SYNC, pin 1 of the switch is grounded (Fig. 7-19). This action grounds the vertical sweep capacitor C16 via CR20, inhibiting the vertical sweep generator. In addition, the

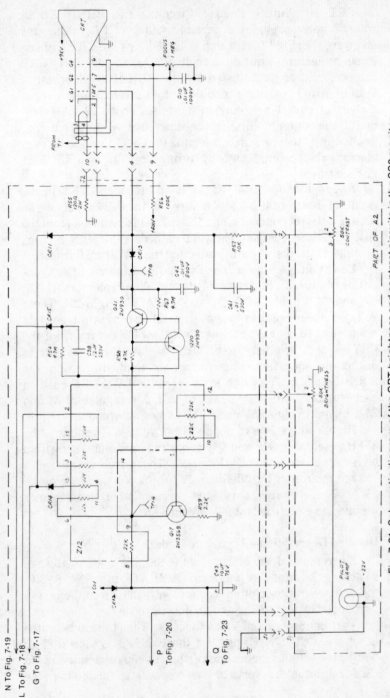

Fig. 7-21. Schematic diagram of the CRT brightness and contrast circuits in the SS2 monitor.

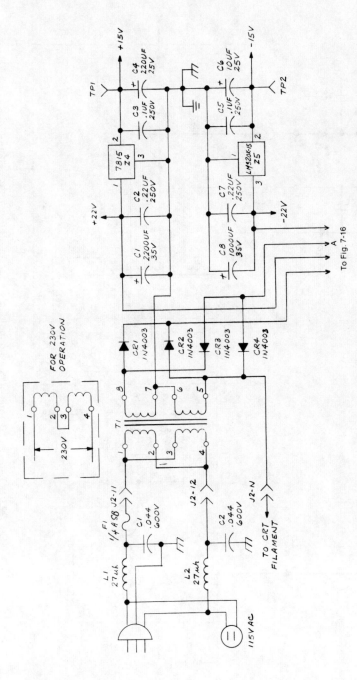

Fig. 7-22. Schematic diagram of the low-voltage power supply in the SS2 monitor.

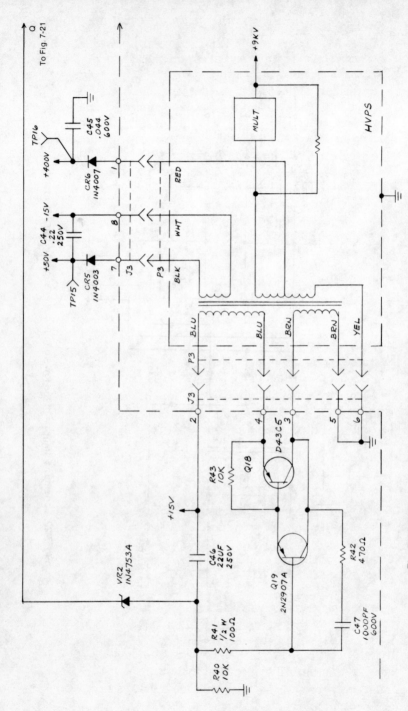

Fig. 7-23. Schematic diagram of the high-voltage power supply in the SS2 monitor.

cathode of diode CR18 is grounded, turning off switch Q5 (Fig. 7-17). As a result, diode CR22 is forward biased, permitting video signals from the output of amplifier Z1B to pass through C12 to vertical sweep amplifier Z3A. This arrangement is consistent with normal oscilloscope operation, that is, the video input rather than the vertical sweep generator controls vertical deflection.

Horizontal Sync Circuit. With pin 1 of the mode switch grounded, CR27 conducts, turning off switch Q4 (Fig. 7-17). This back biases CR25, inhibiting the passage of negative-going horizontal sync pulses from the output of amplifier Z7C to sweep generator Q10. Instead, Q10 is triggered by positive-going pulses from Z7D, coupled through contacts 4 and 5 of the mode switch, and capacitor C19. Since Q10 is triggered by negative-going transitions, a time delay equivalent to the horizontal sync pulse width is introduced. This arrangement permits the horizontal sync pulses to be viewed along with the video.

CRT Brightness and Contrast Control Circuits. Pin 1 of the mode switch also grounds the cathodes of CR11 and CR14 (Fig. 7-21). With CR11 conducting, the CRT control grid is grounded, preventing Z-axis modulation. Turning CR14 on reduces the CRT cathode voltage to a level that still provides normal brightness, compensating for the removal of the DC voltage from the CRT control grid. The mode switch may thus be flipped to either OPERATE or ACCU SYNC without upsetting CRT brightness.

VENUS C1 CAMERA

The C1 camera (Fig. 7-24) is capable of operation as both a SSTV or fast-scan TV camera. This allows an ordinary 525-line home TV to be used as a viewfinder by merely connecting the fast-scan RF output to the set's antenna terminals. SSTV signals are generated using the sampling technique. This process generates one line of slow scan for each fast-scan frame. Other camera features include—microfocus that allows focus from 3 inches to infinity—internal bar generator for perfect monochrome calibration—video polarity reversal for special effects—automatic light-level control on vidicon so that pictures can't be washed out due to excessive light—continuously variable frame length so that much more power for the camera is furnished by its 115V source.

Fig. 7-24. Side view of Venus C1 camera. This camera will operate either slow scan or fast scan.

Operation of the C1 is straightforward and enjoyable, provided you take a few minutes to read the instruction manual. The camera is quickly adjusted with the aid of an ordinary TV, then the mode select is switched to SSTV, and the fun begins. The microfocus control and variable frame rate then give unlimited production capabilities.

General Description

The C1 camera consists of a modified Koyo CCTV camera to which a state-of-the-art SSTV converter has been interfaced. The C1 camera configuration has all the characteristics of both a standard (fast-scan) CCTV camera and an SSTV camera. When operating in fast-scan mode, the composite video output of the camera may be applied to a TV monitor in normal fashion. An RF-VIDEO SELECTOR switch at the rear of the unit also permits the fast-scan output to take the form of a video modulated RF signal that can be applied to the VHF antenna input terminals of a TV receiver so that any TV may be used as a monitor.

In its slow-scan mode, the SSTV signals can be used to form a 128-line image generated at a line rate of 15 Hz (nominal) with a frame rate of 8.5 seconds (nominal). Because of the narrow bandwidth of these AF signals (1200−2300 Hz), the SSTV camera signals can be transmitted over telephone lines, recorded and played back with standard audiotape recorders, or used to modulate SSB transmissions. All of these applications are easily accommodated by the SS2 monitor, the C1 camera's sister unit. A special cable is supplied with the C1 for connection to the SS2.

Converting Fast Scan to SSTV

The fast-to-slow-scan conversion scheme is shown in simplified form in Fig. 7-25. Since slow-scan standards call for a 15 Hz line rate, the frame rate of the fast-scan camera is reduced from 60 Hz to 15 Hz—four times as many fast-scan lines (1048) are generated at the standard 15.75 kHz horizontal line rate. A sampling technique is used to generate a slow-scan image from these fast-scan lines. As shown in Fig. 7-25, slow-scan line 1 is formed by sampling the video signal at the very end of the first fast-scan line, then the end of the second fast-scan line, etc. until the ends of all 1048 lines have been sampled. Thus, a slow-scan *line* is generated from each fast-scan *frame*. Since the fast-scan frame rate is 15 Hz, the slow-scan line rate is established at the proper value. The fast-scan vertical sync pulse (which normally occurs after

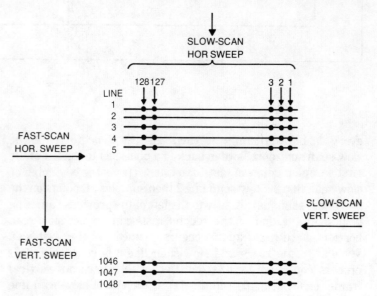

Fig. 7-25. Simplified diagram of fast-to-slow-scan conversion scheme. Each slow-scan line is generated by one fast-scan frame.

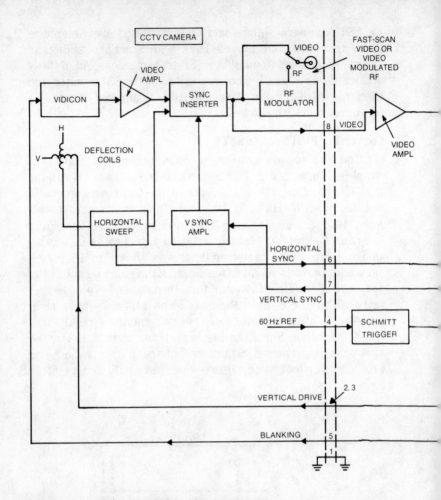

every fast-scan frame) is then used as sync to trace the slow-scan horizontal sweep back to a point just to the left of the first sampled point on fast-scan line 1. This is the beginning of slow-scan line 2. Slow-scan line 2 then samples all points on the 1041 lines, each just slightly to the left of the previous sampling points. At the end of the second fast-scan frame, slow-scan horizontal retrace again occurs (timed to the fast-scan vertical frame rates and slow-scan line 3 is generated. This process continues for the duration of the slow-scan vertical frame, or 8.5 seconds (nominal). At the end of slow-scan line 128, slow-scan vertical retrace occurs, initiating a new slow-scan image. Note that the slow-scan lines sweep at a very

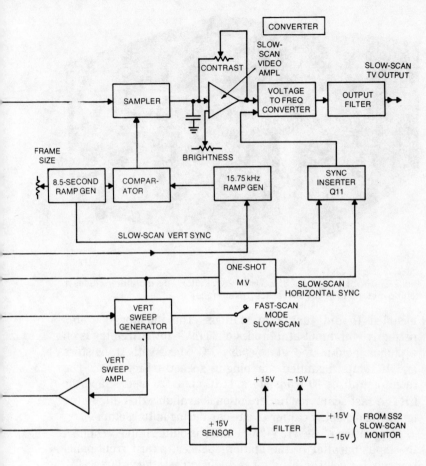

Fig. 7-26. Block diagram of the C1 camera.

slow rate from right to left as the 8.5-second ramp decreases in value. Each slow-scan line actually tilts to the left as it is being scanned. Also note that in order to see the slow-scan image right side up, the camera must be rotated 90° from its fast-scan position. The block diagram for the C1 is shown in Fig. 7-26.

SEEC HCV-2A MONITOR

The HCV-2A (Fig. 7-27) monitor is an attractive and excellent performing unit with some very interesting features. Picture size is 6.25 inches diagonal, and special CRT phosphor mix coupled with a neutral density filter provides monochrome tinted SSTV pictures. A front-panel tuning meter is

Fig. 7-27. Front view of the SEEC HCV-2A monitor. This monitor includes a tuning meter and a manual vertical reset trigger.

included to aid tuning in stations. The monitor operates perfectly with input amplitudes of 20 mV–10V. Circuitry is on a plug-in gold-plated glass-epoxy PC board with transistors and IC chips installed via plug-in sockets (Fig. 7-28). The monitor utilizes 30 transistors, 11 IC chips, 26 diodes, and the CRT. A fast-scan viewfinder option is available that allows the unit to also display camera fast-scan during initial setups.

Operation of the HCV-2A monitor is quite simple—connect the input parallel to the station speaker, adjust front-panel controls for desired brightness and contrast, then view SSTV. The tuning indicator is especially advantageous when you are first learning how to tune SSTV on a short-wave receiver. SEEC also includes three different color CRT filters with each monitor for those who tire of viewing monochrome pictures.

Input Limiter and Discriminator

With reference to the block diagram (Fig. 7-29), the front-panel video selector switch feeds incoming FM audio from an SSTV camera, flying-spot scanner, audiotape recorder, or a radio receiver to limiter amplifier A4. This amplifier, with its associated noise-immunity circuits, removes amplitude variations on the incoming signal over a 60 dB range, producing a clipped square-wave output containing only FM information.

Fig. 7-28. The main PC board in the SEEC HCV-2A monitor is a plug-in type with gold plating on glass epoxy. The circuitry includes 30 transistors, 11 IC chips, and 26 diodes.

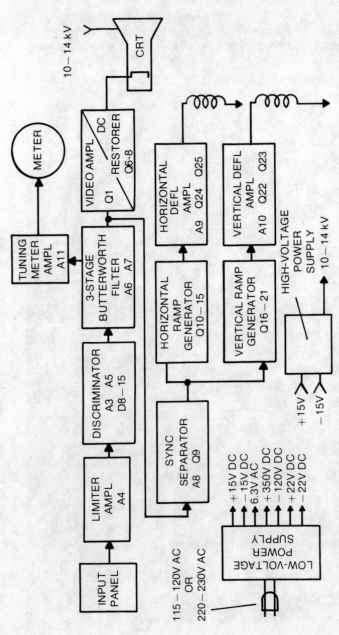

Fig. 7-29. Block diagram of the SEEC HCV-2A monitor. The +15V and −15V DC supplies feed the DC-to-DC high-voltage power supply.

Two tuned FM discriminator circuits possessing the classical S-shaped response, with peaks at 2400 Hz and a low point at 1100 Hz, connect the amplitude limited FM to a 0−1100 Hz baseband AM signal in which amplitude represents brightness on the monitor display CRT. By use of full-wave rectificaton followed by a 3-stage Butterworth filter, ripple is kept to an absolute minimum. The resulting signal is applied to the CRT cathode via a video amplifier, consisting of Q1, Q6, Q7 and Q8.

Sync Separator

In a special designed sync separator, the AM video signal is stripped of the blacker-than-black 1200 Hz horizontal and vertical sync pulses that are inserted in the original picture signal generated in the camera or other SSTV source. It employs a self-biased level-gating circuit which extracts proper sync pulses over a receiver tuning range (or audiotape recorder speed variations) of ±250 Hz. Two remaining circuits then receive the conditioned pulses. One is the PLL circuit consisting of a phase detector, a VCO, and a horizontal deflection amplifier. The incoming sync pulses lock the oscillator—normally free-running at approximately the standard horizontal line rate of 15 Hz--to exact line rate of the incoming video signal. The rate of the horizontal oscillator is varied by the front-panel control, HORIZONTAL HOLD VR9. It may be adjusted by observing the direction of the diagonal black bars, which are very faint and do not interfere with the picture in any way whatsoever. They run across the screen from top to bottom when the picture is not in sync with an incoming signal. With VR9 adjusted to the point where these faint lines disappear, the horizontal line rate will be exactly 15 Hz and ready for an incoming signal. This circuit will also compensate for input variations of ±25% from standard. The output of the horizontal deflection amplifier drives horizontal output transistors Q24 and Q25, which in turn drive the horizontal deflection coils of the CRT deflection yoke with a very linear sawtooth, producing horizontal sweep. The horizontal amplifier output is also applied to a sweep-failure protection circuit which bias off the CRT beam current in the event of a sweep failure, thereby preventing any damage to the CRT phosphor.

The output of the sync separator is also supplied to an integrating gate that *rejects* 5 ms horizontal sync pulses, which occur every 1/15 second and *gates through* the 30 ms vertical pulses that occur every 8.25—8.75 seconds. The pulse triggers the normally free-running vertical oscillator, resetting it to a value corresponding to the top of the display screen and thereby starting the vertical sweep over again for a new picture frame. With no vertical sync pulses present, the oscillator will *free run*, providing a constant raster on the CRT. The vertical oscillator output is applied to a vertical amplifier and complementary transistor pair Q22 and Q23, then to the vertical deflection coils of the deflection yoke. A MANUAL VERTICAL TRIGGER push-button switch is also provided that allows the operator to restart the vertical sweep at any time.

CRT Phosphor and Display

The HCV-2A utilizes a special custom-made CRT that incorporates a mixture of P7, P14, and P21 long-presistence phosphors. The result is a very sharp picture with a white-blue color, rather than the normal P7 yellow-green color. With the addition of a special neutral density filter in front of the CRT, you may view a close-to-black-and-white picture. Also furnished are an amber and a green filter to select your own vision requirements. In most cases the neutral density filter is the only one used.

Tuning Indicator

A 100 μA tuning meter is provided as an aid for tuning in the SSTV signal. It is designed for optimum operation in the RECEIVER position of the video selector switch. It will be noted that the meter will fluctuate slightly whenever a slow-scan signal is not present or whenever an SSTV signal is present but not properly tuned in. This is because meter amplifier A11 is sensing audio signals in the 1200—2300 Hz range in bursts. When a steady SSTV signal is properly tuned in, the meter will indicate a positive and steady red color, usually full red; normal resting position with no signal applied is half-red/half-white; with power off, it will remain all white.

Fast-Scan Viewfinder Option

The fast-scan viewfinder allows the operator to switch from slow scan to fast scan by a flip of the front-panel switch. In the fast-scan position, the monitor's horizontal and vertical rates are changed to match the sampling (fast-scan) rate of the HCV-1B camera. A 4-by-4-inch display is provided on the same CRT used for viewing SSTV and fast-scan pictures that can be viewed while sending slow-scan pictures. The fast-scan picture will also display movement or whatever the camera may be looking at in real time. This feature allows you to adjust focus and set up scenes instantly, whereas normal slow scan requires an 8-second wait period to see what changes have been made.

Miscellaneous Circuits

All supply voltages, +15V DC, −15V DC, 6.3V AC, +250V DC, −120V DC, +22V DC, and −22V DC, are provided within the monitor. In addition a special regulated high-voltage anode voltage, 10−14 kV (adjustable) is provided for the CRT in a separate housing within the monitor cabinet. The primary line voltage for the entire monitor is 115−120V 50-60 Hz or 220−230V 50-60Hz.

An internal telephone-line matching transformer is provided in the monitor connecting (via a rear-panel connector) to the telephone line. This matching transformer meets with all Bell Systems requirements. A fast-scan connector is provided to bring in fast-scan video and sync information from a HCV-1B SSTV camera are standard on all HCV-2A and B models after September 1974, and may be installed on others in service on special request.

SEEC HCV-1B CAMERA

The HCB-1B camera (Fig. 7-30) utilizes a 7735A vidicon operating at fast-scan rates that is output sampled to produce SSTV. Fast-scan video modulates an RF oscillator built in the camera that is tunable to channels 2−6. This feature allows a conventional TV to be used as a viewfinder, provided the horizontal frequency is changed. This change is necessary because fast-scan pictures from the HCV-1B camera consists of 120 lines when viewed on home TV. The HCV-1B camera features a built-in AC power supply (which allows it to be independent of other SSTV gear), video polarity reversal switch, and frame selector of ¼, ½, or ¾ scan. Circuitry is on a plug-in glass-epoxy gold-plated board with transistors and IC chips mounted via plug-in sockets (Fig. 7-31). The camera utilized 48 transistors, 14 IC chips, 26 diodes, and the vidicon. An optional gray-scale generator is also available from SEEC.

Fig. 7-30. The SEEC HCV-1B camera with a 7735A vidicon operating at fast-scan rates.

Operation of the HCV-1B camera is described fully in the instruction manual. Basically, this consists of adjusting camera CONTRAST and BRIGHTNESS controls. The camera fast-scan output is fed to its side, so the results of adjustments are readily noticeable.

Theory of Operation

It is first necessary to provide a means whereby a picture can be taken at a fast-scan rate, then converted to a slow-scan rate for transmission and display as SSTV. This is common procedure used as it permits using a conventional fast-scan TV camera tube for a slow-scan rate system.

Fig. 7-31. Main PC board in the SEEC HCV-1B camera is a plug-in type with gold plating on glass epoxy. This board looks similar to the main PC board in the HCV-2A monitor. The circuitry employs 48 transistors, 14 IC chips, and 26 diodes.

Sampling is a process that takes only a portion of each fast-scan frame. This portion is stored and the time between samples is used for slow-speed transmission of the sample. For SSTV it is convenient to take one line of slow-scan video from each frame of fast-scan video. In a TV camera tube, video is generated as the scanning beam senses the image charge stored on an insulating screen.

Since the sampling process operates the TV camera tube at a fast-scan rate, it is often convenient to provide a local CCTV monitor operating at the camera tube rate. A camera with an RF oscillator operating in the standard TV broadcast area, allows the use of a conventional TV set for monitoring the fast-scan rates. Such a monitor or TV provides a much brighter picture and can also display motion, which is not the case with SSTV. Such a monitor is very useful for camera setup and scene composition.

The sampling function in the camera is accomplished by a circuit that times the occurances of the short samples and another circuit which captures and stores the samples. The sample-and-hold circuit consists of a bidirectional electronic switch and a storage capacitor. The timing operation uses a circuit referred to as a comparator to produce a narrow output pulse at the instant its two inputs are equal.

Video Circuits

With reference to the block diagram (Fig. 7-32), the fast-scan video is generated by the vidicon camera tube. A video amplifier consisting of FET transistor Q11 and operational amplifier A3 increases the signal level to approximately 1.3V over a 300−450 kHz band. Following the video amplifier is keyed DC restorer Q14 and Q15. This circuit clamps the video signal to 0V once per line during the retrace blanking period. DC coupling is maintained thoughout the remaining video circuit.

The control on the front panel of the camera labeled BRIGHTNESS allows you to vary the average gray scale in the transmitted scene. Bias is also added to the video signal for the purpose of insertion of the black level.

Maximum video excursion is limited such that whites do not produce output frequencies above 2300 Hz nor blacks below 1500 Hz. This limiting insures that the video will not go out of the allowed frequency band or interfere with the sync. The video gain is adjustable by way of the *f*/stop or the vidicon target voltage, which is a control on the front panel labeled CONTRAST. By adjustment of these two controls, you can accommodate a wide range of external lighting conditions. The conditioned fast-scan video is applied to a sample-and-hold circuit which consists of 6-diode IC switch A5, a holding capacitor, and a high-impedance follower. The 6-diode switch provides an excellent bidirectional charge flow to the holding capacitor, insuring that each new sample is correct.

Composite sync is added to the video signal by temporarily removing the video signal and substituting a bias of the correct value. Fixed-gain amplifier A6 increases the composite video to the proper level to drive frequency modulator Q19 and Q48. The composite video signal is also coupled to 3-transistor video amplifier Q2, Q3, and Q4, which increases the video level to 2.0V P-P. This 3-transistor video amplifier also feeds RF oscillator Q1, which provides an RF output in the commercial TV broadcast spectrum, channels 2–6. This allows use of a standard TV set for setting up the camera and simultaneously monitoring the fast-scan video rate during slow-scan transmissions.

FM is generated by a controlled current source and unijunction oscillator Q48 operating at twice the output frequency. The correct output frequency is obtained by binary division. The square wave at the correct frequency is band limited in 2-stage Butterworth filter A7 and A8. The FM output level is set to the desired level by a rear-panel control. The FM video output is also switched on and off by way of a transmit/standby switch mounted on the rear panel.

Timing and Sweep Circuits

Vertical and horizontal deflection, blanking, and sync signals are generated in the timing network starting with a 60 Hz oscillator that can free run or be synchronized with the 60 Hz supply frequency. Binary dividers A11 and A12 reduce the 60 Hz to 15 Hz and 1/8 Hz rates for slow scan. The 15 Hz is obtained by dividing 60 Hz by 4. Operation on 50 Hz supply mains requires a division ratio of 3. To change a camera from 60 Hz to 50 Hz operation only requires moving a jumper on the PC board.

The camera may also be operated from a mobile or portable location, even though supply voltages may not be

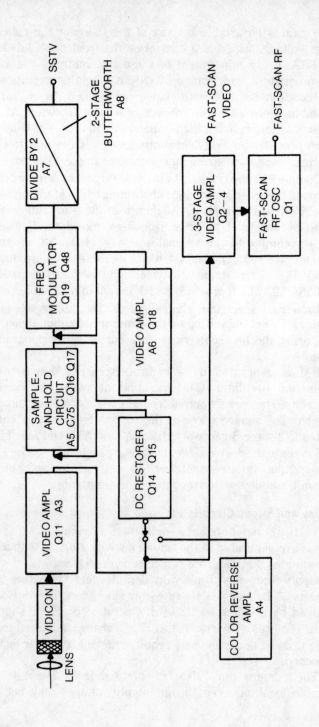

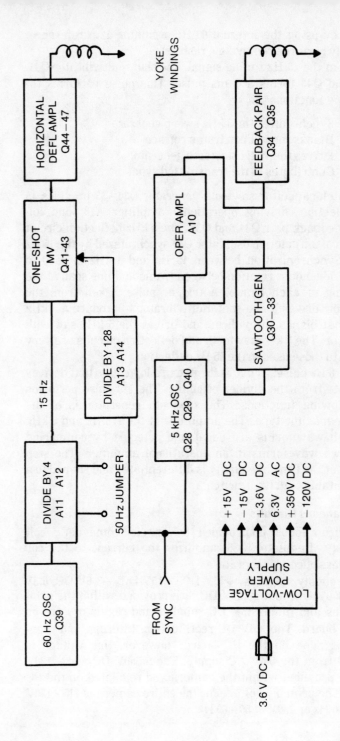

Fig. 7-32. Block diagram of the SEEC HCV-1B camera.

stable, by using the internal 60 Hz oscillator as a reference. This only requires removing a resistor.

From the 15 Hz timing signal, one-shot multivibrator Q41, Q42, and Q43 forms a 5 ms pulse. This pulse performs the following functions:

- Synchronizes the 5 kHz sweep generator
- Blanks the vidicon frame retrace
- Drives the 15 Hz sawtooth generator
- Contributes to the transmitted sync

Switch-capacitor sawtooth generator Q30–33 provides 15 Hz deflection, driving operational amplifier A10, and follower-feedback pair Q34 and Q35. The 5 kHz deflection circuit consists of unijunction oscillator Q40 synchronized by the 15 Hz pulse. Synchronization between 15 Hz and 5 kHz is used to prevent a random relation between frame and line scan at the beginning of each frame. A trigger pulse, taken from the oscillator, drives a one-shot multivibrator to produce a 5 kHz pulse that blanks the vidicon and drives the 5 kHz sawtooth generator. The 5 kHz sawtooth and deflection circuits are very similar to those used in the 15 Hz deflection.

A 1/8 Hz pulse, 30 ms, is formed in a logic gate that derives its inputs from the divider circuitry. The 1/8 pulse performs the following functions: The sawtooth generator is of the switch-capacitor type. The amplitude of the 1/8 Hz and 5 kHz sawtooth waveforms are equal and detected by comparing these two waveforms in an operational amplifier. The step formed at the time of equality is differentiated and shaped to a form suitable to drive a switch.

Miscellaneous Circuits

Vidicon focus is provided by an electromagnetic coil assembly. Regulator Q36 stabilizes the current to the coil regardless of coil temperature.

All supply voltages, +15V DC, −15V DC, +3.6V DC, 6.3V AC, +350V DC, and −120V DC, are provided within the camera. Plus and minus 15V DC supplies and regulators are on the PC board. The 3.6V DC rectification, filtering, and regulation is done on the PC board; however, this voltage is derived from the 6.4V AC supply. The +350V DC and −120V DC are provided within the camera and regulated on the PC board. The primary voltage for the entire camera is 115–120V at 50–60 Hz or 230V at 50–60 Hz.

Focus adjustment is provided in the lens assembly and by a control on the PC board. A front-panel switch provides for reversal of the horizontal deflection coil. This switch can be used with a mirror placed in front of the lens to permit written material lying on the desk to be televised without moving the camera.

A color reversal switch is provided on the rear panel. This allows you to make normal white lettering on a black background appear to be black lettering on a white background when switched to the negative position. This is accomplished by changing the output of video amplifier A3 from positive to negative via color reversal amplifier A4. This feature allows scenes to be televised during very noisy conditions when black letters on a white background stand out better.

Also on the rear panel is a frame selector switch that allows you to program a 1/4 frame, 1/2 frame, and an alternating 1/8 and 3/4 frame. This permits you to use only the necessary screen area. By alternating their use, it is possible to use one of these frame times for one color and the other frame time as the other color. Separate output jacks are provided for fast-scan video, fast-scan RF, and slow-scan FM. An AC power-on switch and indicator are provided as well as an AC circuit breaker. A special jack (Aux. SSTV) for feeding FM video signal to phone patch, (or to related equipment) is provided on rear panel (deleted after July 1974 and available on special request).

A special 4-shade gray-scale generator is provided as an option. The four frequency levels are set at 2300 Hz, 2100 Hz, 1900 Hz, 1700 Hz, but they may be readjusted. A special microfocus (mechanical) adjustment is available as an option. It allows focus down to less than 1 inch with a standard lens. Automatic light control (ALC) is also available as an optional feature. This option provides an AUTO/MANUAL switch to select either ALC or standard operation of the camera. Another option is the fast-scan-viewfinder-sampling-rate output connector that is used to feed the fast-scan sampling rate to a HVC-2A or 2B SSTV monitor that has the fast-scan viewfinder circuitry or the HCV-70-FSVFK (modification kit) installed. This option is standard on all HCB-1B and −1C cameras after July 1974.

Fig. 7-33. The SEEC HCV-3KB SSTV keyboard. This unit produces 35 characters per frame in two formats.

SEEC HCV-3KB SSTV KEYBOARD

This unit (Fig. 7-33) utilizes an ASCII-encoded keyboard and additional digital circuitry to generate SSTV messages in text form. The unit produces 35 characters per frame in a 7-character horizontal and 5-character vertical format or 6 characters per frame in a 3-character horizontal and 2-character vertical format. Special features include: video inversion; 1/4, 1/2, or full-frame scan; fast- and slow-scan outputs; SSTV frequency test keys; and much more. There are 8 basic sections of the keyboard: memory, write clock, read clock, character generator, D-A converter, SSTV VCO, gray-scale generator, and power supply.

SBE SB-1 MTV SSTV MONITOR

The SB-1 MTV monitor (Fig. 7-34) is an attractively enclosed unit that displays sharp SSTV pictures. The built-in cassette tape recorder can record any SSTV picture that is displayed on the monitor screen. This internal recorder also simplifies wiring problems. During transmissions, station ID signals can be run from the tape recorder, while the camera is used for live transmissions. Free-running sweep gives a constant-running raster, and the aluminized CRT displays reasonably bright pictures.

Operation of the SB-1 MTV is straightforward and easy—just connect to the station speaker, adjust brightness and contrast and view SSTV. The internal tape recorder allows instant recording of any special contacts. Likewise, SSTV transmissions can be made from the recorder if desired.

Fig. 7-34. Linear's SBE SSTV equipment. (A) The SB-1 CTV camera with video NORMAL/REVERSE switch. (B) The SB-1 MTV SSTV monitor with built-in cassette tape recorder.

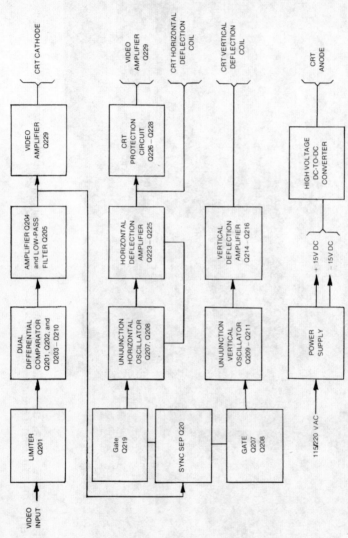

Fig. 7-35. Functional block diagram of the SB-1 MTV monitor. The circuitry includes a CRT protection circuit to burn on the screen in the event of sweep failure. High voltage is developed by a DC-to-DC converter fed by the +15V and −15V DC supplies.

Video Circuitry

Slow-scan video, containing both horizontal and vertical sync received from the internal tape recorder or external source, is limited by Q201 (Fig. 7-35). This reduces undesired noise riding on the composite video signal and reduces any amplitude modulation on the video. The amplitude-limited video is applied to dual differential comparator/discriminator circuit Q202, Q203, and D203−D210 that converts the composite video signal from audio to DC. This composite DC signal is amplified by Q204 and applied to low-pass filter Q205. The output of Q205 is split and applied to the sync separator Q206 and to video amplifier Q229 through CONTRAST control R03. The collector of Q229 is DC coupled to the cathode of the CRT.

Vertical Deflection and Synchronization

Vertical sync is separated from the composite video by the sync separator Q206. Vertical deflection is produced by free-running unijunction vertical oscillator Q209−Q211. This vertical oscillator is synchronized by a vertical sync pulse from integrator-gate Q207 and Q208. An 8.5-second vertical ramp is amplified by vertical deflection amplifier Q214−Q216 and DC coupled to the vertical deflection coil of the CRT.

Horizontal Deflection and Synchronization

Horizontal deflection is produced by free-running unijunction oscillator Q219−Q222. This horizontal oscillator is synchronized with the horizontal sync pulse of the incoming video at Q219. The synchronized 5 ms horizontal deflection ramp is amplified by horizontal deflection amplifier Q223−Q225 and DC coupled to the horizontal deflection coil of the CRT.

CRT Deflection Protection Circuit

The CRT screen is protected against loss of horizontal deflection signal. Video amplifier Q229 is biased off by CRT protection circuit Q226−Q228 whenever loss of horizontal deflection occurs.

Power Supply

The power supply in the SB-1 MTV monitor supplies the following voltages:

- +15V DC regulated
- −15V DC regulated
- +300V DC
- −100V DC
- 16V P-P 60 Hz square wave
- 12 kV DC

All voltages, except the 12 kV, are supplied by transformer T1. The +15V DC is regulated by Q101, Q203, and 6.2V zener D103. The −15V DC is regulated by Q104−Q106 and 6.2V zener D108. The +15V DC and −15V DC also drives the 12 kV DC-to-DC converter Q103, which is the CRT anode voltage.

The +300V DC and −100V DC voltages are unregulated and full-wave rectified.

SBE SB-1 CTV SSTV CAMERA

This SSTV camera (Fig. 7-34) utilizes a vidicon operation at fast-scan rates, which while sampling the output produces SSTV. ALC is included, which compensates for various light conditions. Power for the SB-1 CTV camera is furnished by the SB-1 MTV monitor supply. A 25 mm lens is included with the camera. A video-inversion switch is included for special effects.

Operation of the SB-1 CTV camera is quite simple. Just connect it to the SB-1 MTV monitor—then adjust the BRIGHTNESS, CONTRAST, and BEAM controls. At the station SSTV pictures can easily be recorded on the monitors cassette recorder.

Video Circuitry

The fast-scan video, which is at a 4 kHz line rate and 15 Hz frame rate, is produced by the vidicon tube. The fast-scan video signal is amplified over its 250 kHz range by 2-stage video amplifier Q101 and Q102, followed by operational amplifier/limiter Q103. Bias is added to the video signal at Q105 to control the black level. This level is adjustable by the BRIGHTNESS control located on the front panel of the camera. (See Fig. 7-36.)

The fast-scan video is sampled by bidirectional 6-diode IC switch Q108 to convert it to slow-scan video. (The video is blanked by Q104 while sync is inserted at Q111. Buffer Q112 maintains the proper video level at the input to the VCO.) The output of the VCO is twice the frequency of the slow-scan video

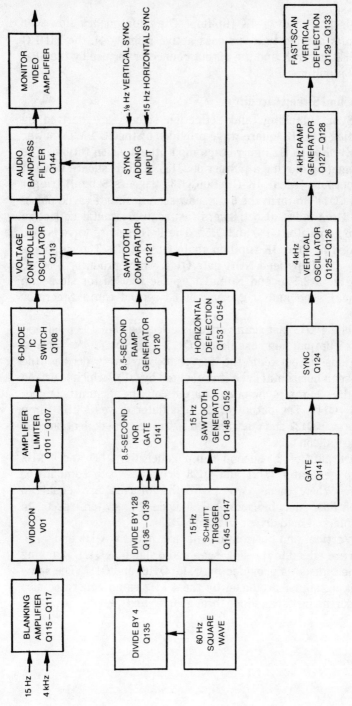

Fig. 7-36. Functional block diagram of the SB-1 CTV camera. This camera uses a 6-diode IC switch to sample the fast-scan video.

and is halved by a JK flip-flop. The *on-frequency* slow-scan video is applied to a low-pass active filter Q144, the VIDEO LEVEL control, and the output connector for use by the SB-1 MTV monitor.

Time and Sweep Circuitry

Sync, blanking, and deflection signals are derived by dividing a 60 Hz square wave provided by the SB-1 MTV power supply. For 50 Hz line voltage input, the division ratio can be changed by moving a jumper on Q135. A 15 Hz square wave at the output of dual JK flip-flop Q135 triggers Schmitt trigger Q145–Q147 to form the 5 ms slow-scan horizontal sync pulse. The 5 ms pulse also triggers fast-scan horizontal deflection ramp generator Q148 and Q152. The 15 Hz square wave is also divided by dual JK flip-flop chain Q136–Q139. The outputs of these flip-flops feed NOR gate Q141, which produces a 66 ms pulse at a 8.5-second rate. This pulse is used for slow-scan vertical sync and triggering of 8.5-second ramp generator Q120.

The 4 kHz fast-scan vertical deflection signal is produced by a unijunction oscillator Q125 and Q126. A random relationship between the fast-scan vertical sync and the fast-scan horizontal sync is prevented by Q124, which is driven by the 5 ms sync-pulse output of the Schmitt trigger Q145–Q147. The output of this oscillator drives 4 kHz ramp generator Q127 and Q128 and fed to the vertical deflection coils of the vidicon tube.

Outputs of 8.5-second ramp generator Q120 and 4 kHz ramp generator Q127 and Q128 drive sawtooth comparator Q121. When these two sawtooth voltages are equal in amplitude, the bidirectional 6-diode IC switch resets to sample-and-hold driver Q122 and Q123.

Vertical and horizontal sync are mixed at Q118 and Q119, then are added to the slow-scan audio signal at Q111. Blanking of the vidicon is provided by Q115, Q116, and Q117. The same blanking signal produced by the 4 kHz ramp generator also blanks the fast-scan video through Q155 and Q104.

Chapter 8
SSTV Satellite Communications

Another area of special interest to many SSTV operators is video communications via amateur satellite. The popular OSCAR series (orbital satellite carrying amateur radio) is ideally suited for these operations. Newcomers to space communications can acquire a wide array of information on satellite work from the American Radio Relay League in Newington, Connecticut or AMSAT in Washington, D.C.

Two of the leading pioneers in OSCAR-SSTV work are Don Miller, W9NTP, and Phil Howlett, WA9UHV. These pioneers established the first two-way video exchanges through OSCAR-6 during the summer of 1974. The following information reflects details supplied to me by WA9UHV and encompasses the full aspect of satellite communication from a personalized viewpoint. This in no way indicates my lack of interest in satellite work—far from it—rather, there are many areas of TV, which makes it difficult to be involved in all areas simultaneously. (See Fig. 8-1.)

Building the Satellite Ground Station

Setting up an amateur SSTV station for satellite communications encompasses more than merely connecting an SSTV source to a 2-meter transmitter and hooking up an SSTV monitor to a 10-meter receiver. This basic setup does provide you with the necessary equipment for satellite work, however, and allows you to view your transmission via

Fig. 8-1. The AMSAT-OSCAR 7 satellite. Photo of the satellite ready for launch. The upper portion of the unit contains solar cells. In the lower monitor display the roll rate was too high.

satellite. The greatest amount of successful satellite work is obtained when each station knows exactly what is happening—when and where to look for SSTV action.

The satellite setup at WA9UHV is typical of most arrangements. This station consists of a Heathkit SB100 transceiver used in conjunction with a Heathkit SB500 10-to-2-meter transverter. Approximate output power of this setup is 40W. A small ventilation fan is placed to cool power-amplifier tubes during SSTV operations. This fan is vital to tube life because reductions in output power from a 40W transverter would seriously impair communications reliability. The 2-meter antenna consists of a pair of 9-element yagis mounted at right angles in a right-hand circular polarization. A pair of antenna rotors in an AZ-EL (azimuth and elevation) mount, approximately 30 feet above ground, follows satellite movements. While this steerable 2-meter antenna is not an absolute necessity, it does make operation much easier. A two-way steerable antenna is especially advantageous during overhead passes of a satellite. The 10-meter antenna consists of two inverted vee's mounted at right angles in a turnstile configuration. Again, right-hand circular polarization is utilized. The 10-meter antenna need not be overly elaborate to perform satisfactorily (Fig. 8-2). The usual crossed-wire dipoles strung between available mounting sources is quite adequate. The inclusion of a high-gain preamplifier with this antenna will substantially improve the signal-to-noise ratio on 10 meters and assist picture copy from the satellite. Likewise, most shortwave receivers can benefit from a touchup of their 10-meter RF coils that were factory tuned for maximum sensitivity between 28.0 MHz and 28.9 MHz.

Tape recording all signals received from the satellite during each pass is advantageous for reviewing and photographing later. Unless you have two tape recorders, the SSTV camera must be used for live transmissions. This arrangement is necessary to free the tape recorder so it can record the 10-meter audio while receiving SSTV pictures.

Satellite Operations

Attempting two-way communication through an OSCAR satellite requires coordination between each party to eliminate hit-or-miss operations. This coordination often includes the use of telephone and low-band links. While hit-or-miss operations are possible, specific planning will eliminate many associated variables. Those of you attempting two-way satellite communication should have your gear set up and operating smoothly *before* a specific launch. Each satellite should be carefully monitored during its first few days aloft. Usually the vehicle roll rate will be very high, causing very deep and prolonged fades. Transmission and reception of SSTV pictures is very difficult during these conditions. After a few days the satellite will stabilize, decreasing the roll rate. Soon after this time SSTV operations will be optimum. Each of the parties attempting satellite work should now compare notes on orbital crossings, acquisition and loss of signals, etc. Coordination via low-band meeting frequencies after each satellite pass are vital during this time. Likewise, scheduling specific 2-meter transmit and 10-meter receive frequencies for side-by-side satellite retransmissions can be established during this time. Each OSCAR satellite has a bandpass well over 100 kHz. Also, antenna tracking must be established during each pass. These operations require specific coordination or a small army of operators for the same function. The most predominant problem in satellite work is low signal-to-noise ratio, producing snow in SSTV pictures. This situation is very similar to the weak signal conditions often experienced on 20 meters. A most successful format for this work is large black letters on a white background. More involved subject matter can be incorporated after satellite roll rate decreases and passes occur between each of the involved stations.

Transmitting your own SSTV picture through the satellite during periods of low activity can also be fruitful. Not only does this allow for evaluation of satellite-related parameters, it can also provide unexpected reception reports from other

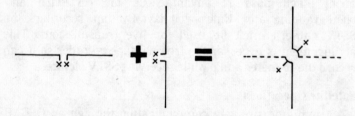

Fig. 8-2. Crossed dipoles for satellite reception. The crossed dipole is constructed from two dipoles fixed at 90°. Feed points are at X. The resulting propagation pattern is omnidirectional — not bidirectional.

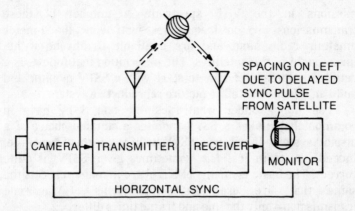

Fig. 8-3. Method of determining the distance that a satellite is from the SSTV station. First, tie the horizontal sync rate from your SSTV camera to the external horizontal sync input of your SSTV monitor. Then transmit any signal to the satellite and pickup the retransmitted signal on your SSTV receiver and monitor. The margin on the left is the time/distance lag between transmit time and receive time.

SSTV stations. There is a good possibility that other stations working via satellite have SSTV capabilities but are listening for activity prior to becoming involved.

Another interesting aspect of singular work via satellite is estimating satellite distances by measuring sync-pulse shifts. Basically, this is accomplished by feeding horizontal sync pulses from the camera to the monitor, then displaying satellite retransmitted pictures on the same monitor (Fig. 8-3). There is a shift of video information on the left edge of each picture; the spacing depends on round-trip distance to the satellite. This arrangement is based on these two facts—radio waves travel 186,000 miles per second and each monitor scan is 66 μs.

The challenge of SSTV work via satellite constitutes a true frontier for you as a technology-pioneering enthusiast. Unlimited experimentation is possible for you if you have ample time and an inventive mind.

WEATHER SATELLITE
PICTURES USING A SSTV MONITOR

The National Oceanic and Atmospheric Administration (NOAA) and the National Environmental Satellite Service (NESS) maintain a series of weather satellites that provide global weather data via real time facsimile picture transmissions in the VHF spectrum. A number of these transmissions are on frequencies just below the 2-meter amateur radio band and are available to anyone within line-of-sight of the satellites. The description that follows is a detailed summary of how to start with a SSTV monitor and build an effective satellite picture reproductive system.

You may wonder what facsimile and SSTV have in common that makes a SSTV monitor a suitable choice of a display system. In essence, SSTV itself is a form of facsimile. Indeed, the term *fast fax*, sometimes given SSTV, is quite correct. The basic elements of SSTV are in many respects the same that are used in some forms of facsimile transmission—only the line and frame times differ.

The satellites covered here are those which utilize narrow band FM (9 kHz deviation) with AM video information on a 2400 Hz subcarrier. That may seem quite restrictive until one goes about the business of keeping up with the present five satellites using this mode. A summary of the currently active (January 1975) domestic satellites and their salient features are shown in Table 8-1.

All of the satellites listed in Table 8-1 use a standard 2400 Hz subcarrier. Maximum amplitude (or deviation) represents white, while minimum—about 4% of maximum—represents black. Line rates vary from 48 lines per minute (LPM) to 240

Table 8-1. Active Domestic Satellites

SATELLITE	FREQUENCY	DATA FORMAT	ORBIT	REMARKS
NOAA-4	137.50	48 LPM visible and infrared spectrums time multiplexed	sun synchronous 115 min period	Real-time readout line scanner
*NOAA-3	137.50	48 LPM visible and infrared spectrums time multiplexed	sun synchronous 115 min period	Real-time readout line scanner
ESSA-8	137.62	240 LPM visible spectrum only, daytime only	sun synchronous 115 min period	Storage vidicon
ATS-1	135.60	240 LPM processed, gridded and mapped NOAA-4 visible and infrared data, global coverage	Geosynchronous 149 W	1200 – 1245 GMT 2200 – 2300 GMT
ATS-3	135.60	240 LPM processed, gridded and mapped NOAA-4 visible and infrared data, global coverage	Geosynchronous 70 W	0420 – 0500 GMT 0900 – 0945 GMT

*NOAA-3 is on operational standby. The VHF transmitter is turned on only when the satellite is not in conflict with NOAA-4. THis occurs. whenever the satellites are separated by more than 23 minutes. Once each month the NOAA satellites transmit on 137.62 MHz for a 1-day period on a nonconflicting basis with ESSA-8.

LPM, depending on the system. The 48 LPM data is continuous in the vertical direction, whereas the 240 LPM data involves vertical frame times of 200 seconds with appropriate vertical interval identification.

The sun-synchronous satellites are in near polar orbits approximately 800 nautical miles above the earth. The time for each revolution around the earth is very near 115 minutes. Each orbit crosses over the equator approximately 28° farther west than the previous pass. The orbital conditions result in daytime passes from north to south in the mornings and nighttime passes from south to north in the early evening. The NOAA and ESSA-8 satellites are in these type orbits. The NOAA satellites transmit real-time weather pictures day and night, by using visible and infrared sensors. ESSA-8 transmits only on the daylight side of the pass. Its sensor is a storage vidicon that takes snapshots at approximately 6-minute intervals.

The data transmitted through the ATS-1 and ATS-3 satellites is referred to as WEFAX and basically consists of the cloud information gathered by the operational NOAA satellite. This data is produced by scan converting, gridding,

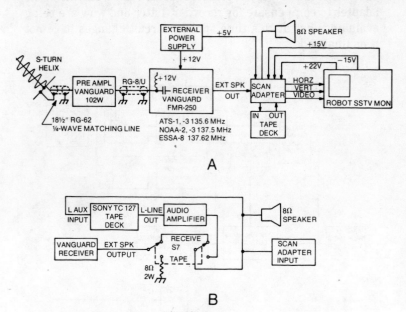

Fig. 8-4. Weather satellite receiving station. (A) Block diagram of receiving gear. (B) Block diagram of audio distribution system (WA7MOV).

and continental outlining to form a series of mozaic pictures. The four daily transmissions, two from each satellite, consist of a series of pictures which cumulatively document the earth's weather conditions for both day and night for each 24 hours. Further information may be found later in this chapter under *Suggested Reading*.

WEATHER SATELLITE RECEIVING STATION

The essential elements for a weather satellite receiving system are a suitable antenna, a VHF FM (30 kHz bandwidth) receiver, a scan adapter, and picture display. The system described here utilizes a Robot 70 SSTV monitor as the display system. The scan adapter developed for this monitor employs a basic approach that could well be applied to most other forms of monitors, oscilliscopes or even monochrome TV sets. The persistance of the P7 CRT is useful but not absolutely necessary especially if the ultimate use is photocopy. As can be seen by the system block diagram, Fig. 8-4, a tape recorder can also be used to record the data for later playback; however, the flutter and wow should be very low if quality equivalent to live copy is required. Some improvements in the adapter to compensate for recorder flutter and wow are being evaluated, but at this time specific circuit changes have not been finalized.

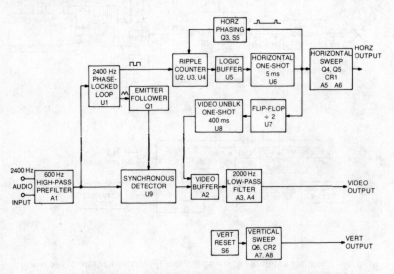

Fig. 8-5. Block diagram of weather receiving adapter (WA7M0V).

A block diagram of the receiving adapter is shown in Fig. 8-5. It identifies the basic video recovery and sweep timing circuit elements. The circuit is basic to most raster-type display systems with the exception of the vertical triggering; it is manual and will be shown later to be a practical choice.

Scan Adapter Circuit Description

The video demodulator is shown in Fig. 8-6. The first stage, A1, is configured as a 600 Hz high-pass filter. This prefilter provides a measure of relief from 60 Hz pickup that somehow often finds its way into systems employing multiple ground returns. From this point the 2400 Hz subcarrier is split into two paths.

One path is to the 565 PLL. The PLL free-running frequency is set near 2400 Hz by R7. A feedback filter is provided by R9 and C7 to remove any amplitude or spurious pulses from shifting the VCO frequency once it has locked onto the subcarrier. Both the square and triangular wave outputs of the PLL are used. The square wave feeds the horizontal sync circuits while the triangular wave provides a carrier reference for the synchronous detector. A second path provides the signal input to the demodulator.

U9, a MC1596/1496 balanced modulator IC, is configured to act as a synchronous detector. This is accomplished by feeding a high-level signal to the carrier port (pins 7 and 8) while maintaining the input to the signal port (pins 1 and 4) within the linear operating region. The original configuration of the detector employed single-ended drive to both the signal and carrier ports as well as a single-ended output. Although satisfactory pictures were produced, a noticeable improvement was noted when the entire circuit was differentially coupled. The differential drive was accomplished by adding amplifier A11 in the carrier circuit and amplifier A10 to the signal path. Q1 acts as a high-impedance buffer for the triangular-wave output of the PLL. The triangular wave is used since it is in phase with the subcarrier signal being applied to the signal port. It is this phase relationship between the carrier and signal port inputs that perform the synchronous detection. Although this technique of demodulation is far more complex than a simple full-wave detector, it does have excellent noise-rejection characteristics not found in the latter. Spurious carriers, such as

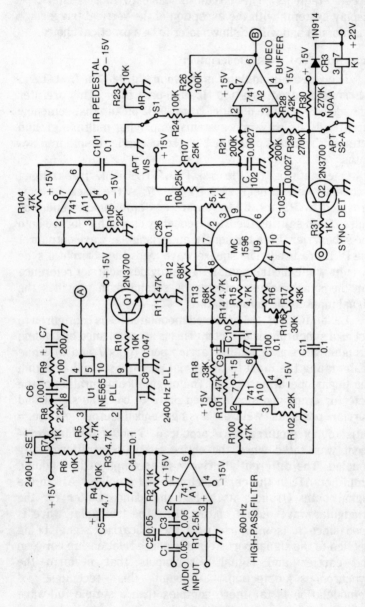

Fig. 8-6. Schematic diagram of the video demodulator for the weather satellite receiving adapter (WA7MOV).

adjacent-channel aircraft transmissions, have far less effect on the picture reproduction than they would using non-synchronous detectors.

Although the inputs are AC coupled, there can be sufficient DC offset at the two input terminals. For this condition the circuit acts like a modulator and results in what appears to be a 2400 Hz feedthrough which is independent of the signal amplitude. Fortunately a simple remedy exists. The input ports are balanced by R106 to null any DC offset. This is accomplished by observing the output of A2 when no signal is applied to the adapter input. The adjustment of R106 should bring the 2400 Hz output right down to almost zero.

The output of the detector is fed to amplifier A2, where it is summed with two bias signals. The use of these bias signals will be discussed in more detail in the section concerning operation. A two-stage cascaded active filter, shown in Fig. 8-7, completes the removal of subcarrier products, leaving a very clean 2V video signal for the monitor. Coupling of the adapter video into the monitor is accomplished with relay K4.

Unlike the SSTV mode, the video level on the CRT is a function of both the monitor *contrast* control and the receiver *volume* control. This does not mean that you must *ride gain* on the signals since the basic transmission is FM and signal limiting takes place in the receiver before audio detection. A receiver *volume* setting, which yields 2−2.5V of video at the output of A4, will result in an average SSTV monitor *contrast* setting. Two additional functions, shown in Fig. 8-7, are worth noting before proceeding further.

Capacitor C19 is connected in parallel with the monitor C34 when displaying weather pictures. This is needed to prevent the monitor sweep-sensing circuits from cutting off the video due to the lower sweep rates used for weather data.

Oscilloscope monitoring of the adapter video output has a very practical value in insuring uniformity from picture to picture and from day to day. In addition, the operator can observe the *on-stream* video without actually seeing the CRT as is often the case when photocopying. Amplifier A9 is intended to give a buffered interface between the scope and video. All versions of the adapter built to date have not required A9; instead the video is monitored at the output of A4 and the *contrast* set and left at predetermined positions. Direct monitoring of the *contrast* wiper will increase line impedances and possible pickup problems.

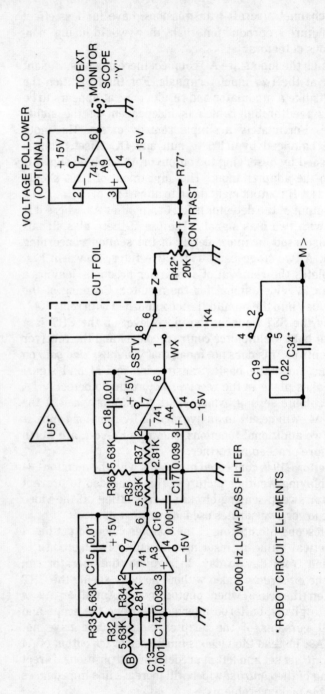

Fig. 8-7. Schematic diagram of the video filter for the weather satellite receiving adapter (WA7MOV).

282

Horizontal drive to the scope can be obtained by direct coupling to the horizontal yoke current-sampling resistor or external triggering from the adapter horizontal sync pulse. The former has the advantage of not needing to change oscilloscope settings for the different sweep rates, WX or SSTV.

The second basic function of the adapter is to provide the necessary horizontal and vertical sweep drive. The vertical sweep triggering is totally manual and will be discussed in the section on operation. The fundamental basis of the horizontal sweep control is derived from the 2400 Hz subcarrier. As it turns out, each of the various satellites provide horizontal sync pulses—unfortunately they are not all the same. To illustrate, the WEFAX and ESSA-8 have the same horizontal line rate, 240 LPM, but each has a different type of sync pulse. ESSA-8 has a white sync pulse during picture transmission—none during the period between pictures and picture start tone—and a black pulse during the 5-second phasing interval. WEFAX is just a little different. A white pulse is sent at the start of each active picture line; however, every other line the pulse is preceded by an alternating black and white pulse about twice as long as the sync pulse. The picture start tone (two sequential tones) has no sync pulses. A 25-second phasing interval has black sync pulses on every other line. The NOAA series use a 7-cycle burst of 300 Hz to signify the start of each of the two data spectrums being transmitted. Which one is which is determined by the level of the pre-earth scan which follows the sync pulse; black is visible and white is infrared. Rather than construct the several special-purpose circuits needed to recover the various types of sync pulses, a counter and your eye can easily accommodate all forms.

The 2400 Hz subcarrier is a form of master clock, and all the satellites have their horizontal timing locked to it; you only need to recover the clock and count it. The 240 LPM systems require 600 clock pulses per line while the 48 LPM systems use 3000. The source for the clock is the PLL, which is also used in the video demodulator. A selectable ripple counter provides the rest.

Figure 8-8 shows the circuitry used to generate the two basic horizontal time bases. The square-wave output from the PLL is coupled to a frequency divider, consisting of U2, U3, and U4, by capacitor C23 and pull-down resistor R67. The AC

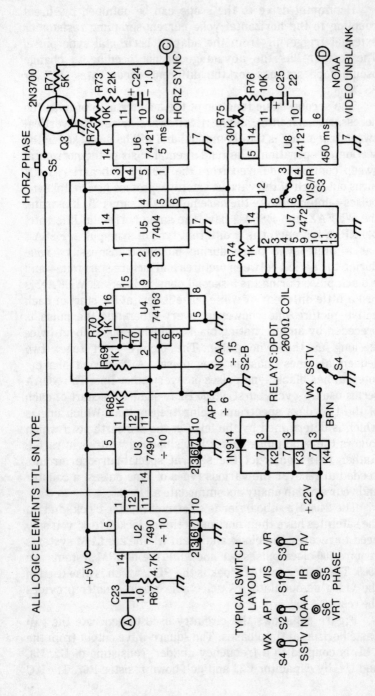

Fig. 8-8. Schematic diagram of the counter and control for the weather satellite receiving adapter (WA7M0V).

coupling is needed to remove the DC component of the square wave that results when the PLL is operated from a single supply. U2 and U3 are connected in cascade yielding a divide by 100. U4 functions as a divide by 6 or 15, depending on the position of S2-B. U5 couples the counter output to 5 ms one-shot multivibrator U6 that, in turn, becomes the trigger pulse to reset the horizontal sweep. The output of the one-shot multivibrator is also coupled back to U2 via switch S5 (a normally open pushbutton switch). Closing S5 alters the counter by stopping the counting for 5 ms each line as long as the switch is closed. This increases the count period to slightly longer than that of the data. The result is to shift the picture on the CRT to the left. Your eye will tell you when the sync pulse is properly aligned at the side of the picture at which time S5 is released. The horizontal will remain in registration as long as the PLL remains in lock (except between pictures for ESSA-8 as noted earlier). Do not entertain thoughts of shifting the PLL for phasing, the video will be lost, hence no definition of the sync pulse.

R72 sets the length of the internal trigger pulse, and anything from 3 to 10 ms can be used. A long pulse will not interfere with the picture but will cause a large picture jump when phasing.

U7 and U8 are used only when processing the NOAA 48 LPM time-multiplexed pictures. The basic horizontal trigger pulses from U6 cause the display to be scanned at 96 LPM, or twice the line rate. U7 divides this by a factor of two and triggers video unblanker U8. In this fashion, only every other scan is enabled. S3 is a selector switch to allow the operation to choose which data field, visible or infrared, to display. R75 is used to set the unblank period at approximately 400 ms. This allows time for a scan to include a sync burst, pre-earth, earth, and post-earth elements. All other data (telemetry and alternate field) are blanked to prevent tube side flash or overwriting. The manual-select switches prevent blanking from being applied to the 240 LPM data for which none is desired.

The horizontal and vertical sweep generators, shown in Fig. 8-9, are equivalent circuits except for the values of the capacitors and current source resistors. Both circuits have free-running times that exceed the triggering times, thus maintaining a reasonable duty cycle on the power-handling

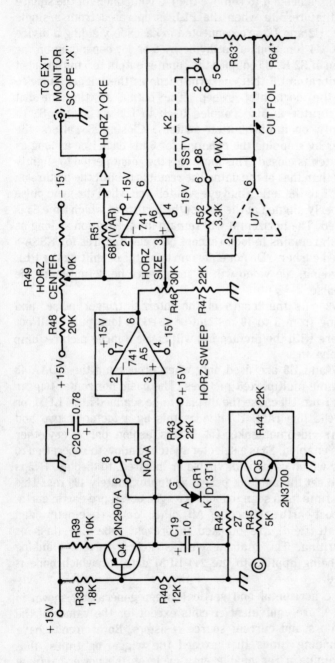

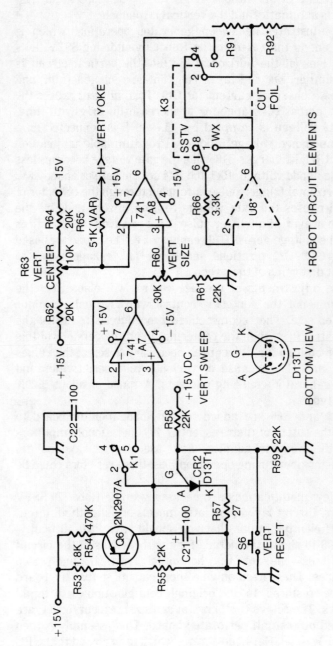

Fig. 8-9. Schematic diagram of the horizontal and vertical deflections circuits for the weather satellite receiving adapter (WA7MOV).

components even without triggering. The horizontal triggering is automatic, and the vertical is manual.

To illustrate the sweep generator operation, which is equivalent to that used in the Robot monitor for SSTV, let's take a look at the horizontal circuit—the vertical circuit is quite similar. Q4 acts as a constant-current generator and supplies a charging current for C19. This, in turn, causes the voltage across the capacitor to increase linearly with time. The C19 voltage is sampled by A5, which is connected as a high-impedance voltage follower. Programmable unijunction (PUT) CR1 discharges C19 when the gate voltage becomes less than the anode voltage. R43 and R44 set the gate at about 7.5V. The horizontal trigger pulse is coupled through the collector of Q5 and causes the gate voltage to drop to zero and fire the PUT; however, if a trigger pulse is not received the PUT fires when the voltage across C19 reaches 7.5V. The capacitors used in the C19−C22 positions should be low-leakage types to prevent distortion of the sweep.

The adjustments of the size control will also affect the centering, but the centering control will have only a minor affect on size. The recommended procedure is to adjust the size control first, then the centering. In all the versions of this adapter built so far, it has been necessary to select the values of R51 and R65 because of variable yoke sensitivities and sweep generator charging currents. A range of about ±20K seems typical.

R52 and R66 are added to the Robot monitor board to clamp the output of their respective 709 operational amplifiers when the weather adapter is in use. If you have no apprehension about using open-loop 709s, these resistors could be left out.

A few modifications are necessary to the Robot 70 SSTV monitor. Circuit layouts on later models are slightly different but sufficiently similar that you should have little difficulty. Figure 8-10 shows the details for the three cuts of the circuit board foils—one for the horizontal, vertical, and video interfaces. These need only be short cutouts so that the board may be restored to its original configuration with small jumpers. The relays (half crystal can size), except for K1, are mounted on a small perforated board. This assembly is then set on 1 1/2-inch flat-head screws which are epoxied to the PC board between components. Leads to the adapter board, signal

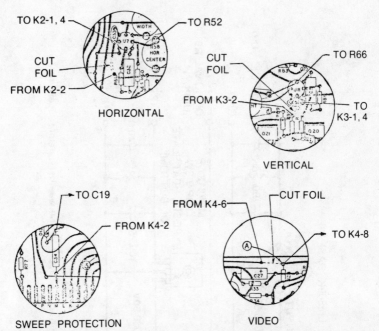

Fig. 8-10. Modifications to the circuitry of the Robot monitor PC board when used in conjunction with the weather satellite receiving adapter (WA7MOV).

and power, are terminated on the relay board and grouped in a flat bundle. The wire bundle is brought out of the monitor by passing it over the top of the rear panel but underneath the top cover. This way no additional holes are needed and the monitor and adapter circuit boards can be readily removed. S4, the SSTV/WX switch, can be mounted on the front panel (small toggle switch) near the top and just to the right of the *brightness* control. No problem has been encountered with the additional power drain on the monitor power supply. The relays used are rated at 28V DC at 10 mA. A power supply suitable for the adapter at 5V DC and VHF receiver at 12V DC is shown in Fig. 8-11.

The adapter can be built using ¼W resistors; 5% tolerance resistors should be used in the circuits of A1, A2, A3, A4, A10, A11, and U9. Capacitors in the sweep circuits as well as the active filters should be low leakage. Use by-pass capacitors, both 22 μF at 20V DC electrolytics and 0.01 μF ceramics, on various points along the +15, −15, and +5V DC power

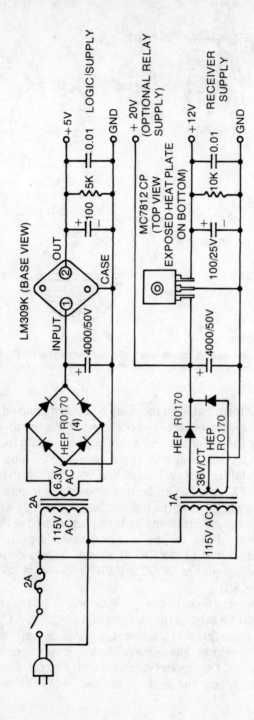

Fig. 8-11. Schematic diagram of the logic and receiver power supply for the weather satellite receiving adapter (WA7MOV).

distribution wiring. A large copper wire ground bus is also used to reduce circuit cross talk.

Operation

This is not a display system from which the weather pictures can be viewed directly. The WEFAX pictures, which are sent in an ordered sequence, can be viewed in a darkened room. The CRT persistance is far from long enough for a full 3-minute frame to be seen in total.

The first operational task is to set the adapter size and centering controls. The *horizontal size* should be set with the PLL synced to a satellite signal, live or recorded, or by setting the free-running frequency as close to 2400 Hz as possible. With S2 set in the APT position, set the *width* so that the raster is scanned to within ¼-inch of the sides of the bezel. The vertical adjustment will take a bit of time in that a frame is 200 seconds long. Press manual *reset* switch S6. While holding it down adjust the *vertical centering* so that the trace is 1/4-inch below the top of the viewable screen. Release the switch and note the position of the scan line 200 seconds later. If it is not 1/4-inch from the bottom, adjust the *vertical size* and repeat again holding the *reset* while setting the scan line to the top with the *centering* control. No input signal is needed for the *vertical size* and *centering* controls.

Switch S2 to the NOAA position, and manually reset the vertical. The horizontal trace should be visible to the very right edge; if not, adjust video unblank control R75 until it is. The time for a vertical scan now will be in the order of 9 minutes, depending on the tolerance of the additional sweep capacitor C22. This is the exposure time for one NOAA picture.

The display is now ready for operation. Remove the amber filter and observe the screen. Set the receiver *volume* control so that the satellite signal produces about 2V P-P output from the adapter. The optimum settings of the *brightness* and *contrast* will come with experience. A good rule of thumb is that the black picture elements just extinguishes the yellow and the white is just short of causing the line to bloom. Gray-scale frames are sent at the beginning of each WEFAX broadcast and make an excellent test of system setup and linearity.

Regarding vertical triggering, numerous vertical-interval identifications are used. For WEFAX, the period between

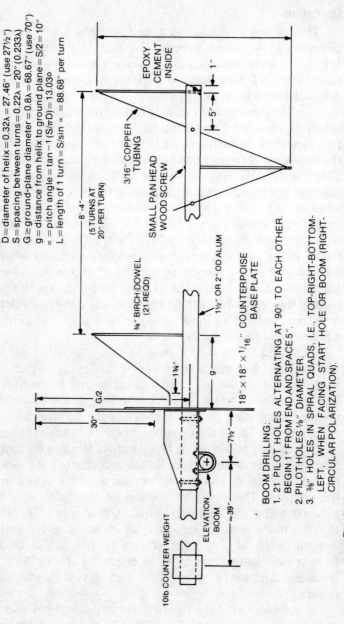

Fig. 8-12. Construction of helix antenna for the weather satellite receiving station (WA7MOV).

pictures consists of 8 seconds of 300 Hz, followed by 8 seconds of 150 Hz, followed by 25 seconds of white with black sync pulses every other line. ESSA-8 uses 3 seconds of 300 Hz, followed by 5 seconds of white with black sync pulses on every line. ESSA-8 pictures are transmitted once every 6 minutes, give or take a few seconds. The time between the end of the last picture and the start of the next is about 2 minutes 40 seconds. This time period is filled by a white carrier level with no sync information since there is no horizontal sync continuity from picture to picture on ESSA-8. The NOAA satellites produce a picture which is continuous, so no vertical framing is used. Just start the picture at the top, and 9 minutes later it will be through. So again, your eyes and ears are used as the vertical sensor. The art of manual triggering is accomplished with just a little practice.

Antenna

Figures 8-12 through 8-16 give the construction details for a 5-turn helix and an AZ-EL mount. This does not represent the last or latest word in satellite antennas, but it works. It accomplishes two things: First of all, though the signal from the satellite is for the most part linearily polarized, the polarization angle is continually shifting. The 3 dB loss of the circular-polarized helix compared to a properly oriented linear array is offset by the lack of any polarization fades. Dimensions and amount of counterweight are approximate and vary with the types of materials used. Covering the counterpoise with ¼-inch hardware cloth improved the reduction of side and back response. The second feature of the helix is that your neighbors will believe you when you say it is for satellite work. Try your 2-meter FM rig on it; it works.

Photocopy

In that it is not convenient to view the pictures directly, the camera affords the most practical means to transform the CRT image into a lasting picture (Fig. 8-17). The possibilities for photocopy are as limitless as there are varieties of cameras. In terms of *instant* weather pictues, the Polaroid is excellent. There are some problems which must be overcome however—a minimum exposure time is 200 seconds for WEFAX and 9 minutes for NOAA. The electric-eye camera must hold the shutter open for that period of time. This is

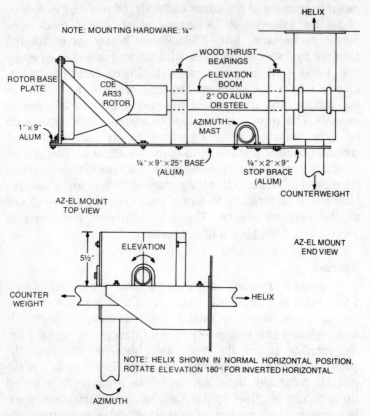

Fig. 8-13. Helix antenna mounting details for the weather satellite receiving station (WA7MOV).

accomplished by covering the sensor lens. Be sure to have a supply of shutter batteries on hand since the shutter draws current as long as it is open. Another problem is the need for closeup lens. A system that works well is built around a Polaroid 110A *Pathfinder* camera.

This is one of the older models which can be obtained second hand through many camera shops. It has a manual lens (hence no batteries) and provisions for a cable released shutter control. It uses roll-type film which can sometimes be a disadvantage. The basic lens on the camera is 127 mm to which two closeup lenses, 3× and 1×, are added to allow copy of the CRT. The image size on the CRT is about 4 1/2 inches square, and the camera is mounted directly in front of it with the camera lens 9 inches from the tube. A plastic hood extends toward the camera from the bezel. A black cloth covers the hood and camera to form a light-tight system.

Polariod 42 film is used. It is a fairly slow, ASA 200, and fine grain film. The faster films can also be used but, setup is more critical. With this film and a lens setting of $f/22$, a display brightness as discussed earlier produces a full-range picture. Good negative densities with the same picture level have been obtained using *Plus X* at $f/8$.

Suggested Reading

The following is a list of U.S. Government publications which provide excellent background information for anyone

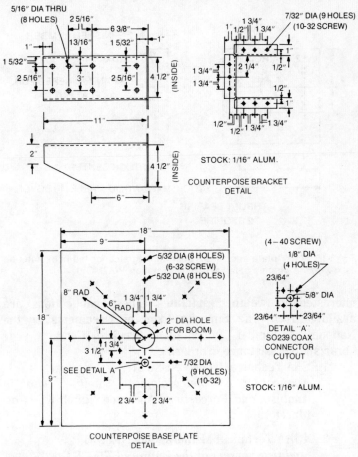

Fig. 8-14. Counterpoise bracket and base plate detailed diagram for the helix antenna used with the weather receiving station (WA7MOV).

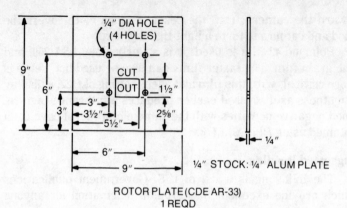

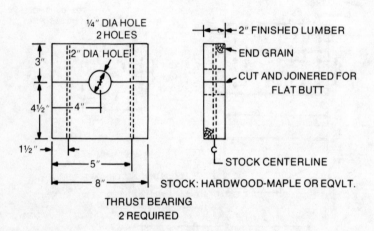

Fig. 8-15. Rotor plate and thrust bearing diagrams for the helix antenna used with the weather satellite receiving station (WA7MOV).

interested in weather satellites. These publications are available from the Superintendent of Documents or the National Technical Information Service. Contact your librarian for assistance in ordering.

1. ESSA Technical Report NESC 51

    Application of Meteorological Satellite Data in Analysis and Forcasting; Anderson, Ralph K., *et al*, March 1974

2. NOAA Technical Memorandum NESS 35

    Modified version of the Improved TIROS Operational Satellite (ITOS D-G); A. Schwalb, April 1972

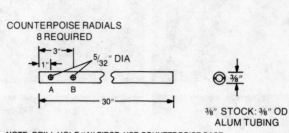

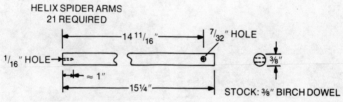

Fig. 8-16. Construction details for the rotor base plate brace, counterpoise radials, and helix spider arms for the helix antenna used with the weather satellite receiving station (WA7MOV).

3. NOAA Technical Memorandum NESS 53
   Catalog of Operation Satellite Products; Eugene R. Hoppe, Abraham L. Ruiz, March 1974

4. NOAA Technical Memorandum NESS 54
   A Method of Converting the SMS/GOES WEFAX Frequency (1691 MHz) to the Existing APT/WEFAX Frequency (137 MHz); John J. Nagle, April 1974

5. NOAA Technical Memorandum NESS 60
   The Operation of the NOAA Polar Satellite System; Joseph J. Fortura, Larry H. Hambrick, November 1974

6. NOAA Technical Report NMFS SSRF—669
   Subpoint Prediction for Direct Readout Meteorological Satellites; L.E. Eber, August 1973

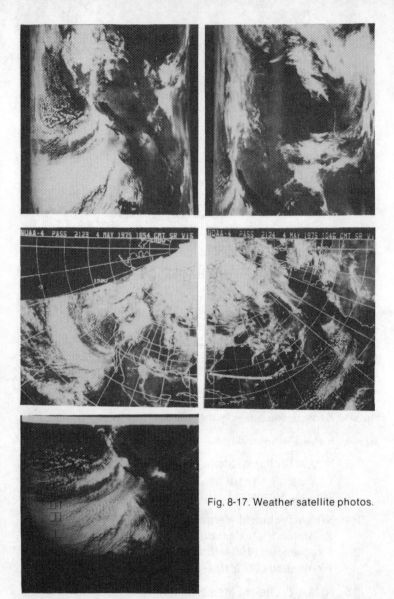

Fig. 8-17. Weather satellite photos.

Further information regarding the satellite systems and status can be obtained by contacting:

APT Coordinator, S122x1
U.S. Department of Commerce
National Oceanic and Atmospheric Administration
National Environmental Satellite Service
Washington, D.C. 20233

Glossary

accelerating electrode—Electrode in CRT for increasing the velocity of electron beam. Usually operated at high-positive potential. Also called accelerator.

aluminized screen—A CRT screen with a thin coating of aluminum on the back of its phosphor layer. Electrons penetrate the aluminum and activate the screen. The aluminum acts as a reflector thus improving picture brilliance.

analog—Representation of quantities by physical variables, like voltage, resistance, etc.

aperture—Another name for the dot produced on a CRT by the electron gun.

astigmatism—A CRT variable which causes the dot to be focused differently for various areas on the screen.

aspect ratio—Ratio of picture width to picture height. This ratio is presently 4:3 for fast-scan TV and 1:1 for SSTV.

blanking level—TV level which separates video signal from sync signal.

burned-in image—Image which persists on camera pickup tube after image is moved from front of tube.

black frequency—In SSTV it is set at 1500 Hz.

capacity effect—Ability of a camera pickup tube to store information in a capacitive charge form.

dark current—The current through a light-sensitive device under total darkness.

definition—The amount of fine detail in a TV picture. High definition indicates a very fine detail picture.

divergence—Spreading of an electron beam. This causes defocusing in CCT.

dynamic operating condition—Relates to the applied signal state of unit.

electromagnetic deflection—Deflection of a CRT beam by means of coils around the tube's neck.

electromagnetic deflection—Deflection of a CRT beam by means of plates mounted inside of a tube. This process must be accomplished during manufacturing.

electrostatic focus—Accomplished by a focus electrode inside a CRT. This electrode produces a static focus field.

facsimile—A method of communicating printed material like maps or photographs.

field—One complete scan of a TV picture. Television broadcast stations in the United States utilize two fields per frame to produce interlaced scanning. Field rate is 1/60 second.

frame—One complete TV picture. In commercial TV, this consists of two fields. Frame rate is 1/30 second. Each complete SSTV picture is considered a frame, consisting of an 8-second period.

fluorescence—Light emitted from a CRT screen when excited by the electron scanning beam.

gate circuit—A circuit which allows signals to pass only when a gating pulse is present.

ghost—Dim image to right of object on TV screen. Phenomena usually caused by multipath propagation.

herringbone interference—Interference to video caused by beating signals with video frequencies.

hum modulation—When SSTV scanning lines cause four slight "waves" in picture (60 Hz divided by 4 equals 15 Hz horizontal scanning rate).

intensity control—Same as brightness control.

keystone raster—Picture with top (or bottom) of different sizes. Raster is trapezodial rather than square or rectangular.

luminance—Brightness of CRT screen when excited by electron beam.

monitor—In SSTV, the actual unit which displays the received picture.

one-shot multivibrator—Same as monostable multivibrator.

peak white—In SSTV it is set at 2300 Hz.

phosphorescence—The afterglow of a CRT.

persistance—Length of time a picture is visible on CRT after excitation.

picture elements—The smallest quantity that a TV screen can be divided.

read cycle—Time when information is extracted from digital memory.

resolution—Same as definition; relates to the maximum number of picture elements which can be resolved.

snow—White dots in video of a TV picture.

static operating conditions—Relates to a no-signal condition for one particular unit.

subcarrier—A carrier which is used to modulate another carrier.

sync frequency—In SSTV it is set at 1200 Hz.

target—Photosensitive area on end of TV pickup tube. Essentially, this element does for video what microphone does for audio.

triggered-sweep monitor—Requires a trigger pulse to scan each horizontal line—no sync pulses—no raster (also called driven-sweep system). The other SSTV system is free-running sweep which syncs to incoming pulses.

turnstile—Antenna composed of two dipoles mounted at right angles and fed 90° out of phase—advantageous in satellite work.

ultor—High-voltage connector for CRT.

voting system—For SSTV: A digitalized technique for processing a good quality picture from three poor quality pictures.

white frequency—In SSTV it is set at 2300 Hz.

write cycle—Relates to time when information is loaded into a digital memory.

X-rays—A form of electromagnetic radiation often resulting from excessive anode voltage on a CRT. These rays are above ultraviolet in frequency.

X-axis—Horizontal deflection of beam in CRT.

Y-axis—Vertical deflection of beam in CRT.

Z-axis—Intensity modulation of beam in CRT.

Index

A

Accu sync mode	232
detailed description	241
A-D converter	157
Alignment	166
Amplifier, video	150, 223
Applications, MXV-200	168

B

Binary encoder	
design of	113
use of	112
Buffers	131

C

Cameras	36, 52
adjustment to monitor	55
electron beam	55
film	58
for SBE SB-1 CTV SSTV	268
illumination adjustment	56
lenses	56
sampling	40
SEEC HCV-1B	256
setting up	52
types of	56
Chain	
timing and sweep	218
video	216
Circuitry, video for SBE SB-1 CTV)	268
Circuits	
color	183
CRT brightness and contrast controls	245
CRT brightness control	241
Circuit	
CRT deflection	267
horizontal sync	245
miscellaneous	255
miscellaneous for SEEC HCV-1B	262
scan adapter description	279
time and sweep for SSB SB-1 CTV	270
timing and sweep for SEEC HCV-1B	259
video	241, 267
video for SEEC HCV-1B	258
video processing	237
Clock	
generator	159
slow	127
Clocking	127
two-phase	131
Color	
circuits	183
receiving system	181
Construction, MXV-200	166
Conversion	
digital fast-to-slow	44
digital slow-to-fast	45
digital slow-, fast-scan	44
direct scan	42
to fast scan TV	247
Converter	
A-D	157
D-A	139, 162
PC boards	141
CRT	
brightness control circuits	241, 245
contrast control circuits	245
deflection circuit	267
phosphor and display	254

D

D-A converter	139, 162
Decimal-binary equivalents	109
Deflection	92
electrostatic vs electromagnetic	92
horizontal	267
vertical	267
Digital	
camera, design of	116
fast- to slow-scan conversion	44

scan converter, WOLMD	118
slow- to fast-scan conversion	45
TV, elementary principles	107
Direct scan conversion	42
Discriminator and input limiter	210

F

Fast-scan timing	127
Flying-spot scanners	34
FM-to-gray-code	124
Frame sequential	177

G

Gear, buying considerations	191
Generator, clock	159

H

HCV-2A monitor	249
Horizontal	
deflection	267
sweep	223
sync	234
sync circuit	245
synchronization	267

I

IC chips	123
Identification, of lines	153
Indicator, tuning	254
Input limiter and discriminator	210, 250

J

Jiggled lines	75

K

Keyboard, for SEEC HCV-3KB	264

L

LED devices	125
Lens	
characteristics	219
zoom	221
Line	
identification	153
jiggled	75
skewed	75
Low voltage supply power	241

M

Main memory	137
Mathematics, relationship between channels and color signals	176
Memory	
color transmission	178
digital	157
Mode	
accu sync	232
operate	232

Monitors	
description	30
Monitor	
for SBE SB-1 MTV SSTV	264
HCV-2A	249
homemade	33
RF feedback	103
setting up	50
theory of operation WOLMD	100
tuneup and alignment	93
WOLMD	95
WOLMD tuning	101
Monitor circuits	
description	85
design simplicity	83
interference protection	81
power	93
quality	81
reliability	81
W9LU0 Mark II	85
Multipath propagation	73
MXV-200	
applications	168
construction	166
parts list	170
scanverter	149

O

Operate mode	232
detailed description	232
Operation	
weather satellite station	291
with telephone line	214
Options, loading and transmitting color	178

P

Parts list, MXV-200	170
Photocopy, of satellite weather pictures	293
Picture framing, rules of balance	72
exceptions	72
Power supply	135, 164, 267, 212
Program production	67

R

RAM memories	115
Readings, weather satellite	293
Receiving system, color	181
RF feedback	103
Robot	
SSTV monitor	191
60 and 61 viewfinders	222
80A SSTV camera	215
300 SSTV scan converter	225
400 converter	225

S

Sampling camera	40
uses	39
Satellite	
ground station	270
operations	272

photocopy	293
weather	274
weather antenna	293
weather, readings on	293
weather station operation	291
SBE SB-1	
CT SSTV camera	268
MTV SSTV monitor	264
Scan	
adapter	279
conversions, convert tube	42
conversions, digital electronic	42
conversion project, sample	120
Scanners, flying-spot	34
Scanverter, MXV-200	143
SEEC HCV-1B	
camera	255
camera, theory of operation	256
miscellaneous circuits	262
SEEC HCV-1B	
timing and sweep circuits	259
video circuits	258
SEEC HCV-3KB keyboard	264
Separator, sync	151, 253
Shift-register tester	141
Skewed lines	75
Slow	
clock	127
scan color converter, W9NTP	173
scan sync detection	126
SSB SB-1 CTV	
time and sweep circuit	270
video circuitry	268
SSTV	
around the world	21
cameras	52
definition	12
description of use	13
DX	68
frequency filter	163
frequency modulator	163
historical view	9
in space	28
in the home	15
loss in horizontal sync	78
monitors	50
swithout monitor	73
noise in video	78
picture analysis	73
picture framing	71
power requirements	69
procedures of operation	65
standards	14
station, starting	48
studio, lighting	61
setting up	61
use on the oceans	24
Station	
ground for satellite	270
weather satellite receiving	276
Supply power, low voltage	241

Sweep	
horizontal	223
vertical	223, 241
Sync separator	151, 212, 253
Sychronization	
horizontal	267
vertical	267

T

Tape recorders	60
Taping, procedures	60
Test point waveforms	142
Timing and sweep chain	218
Transmissions, live	71
Triggered-sweep system	86
Tuning	
characteristics	213
indicator	254
Two-phase clocking	131

U

Update controls	135

V

Venus C1 camera	245
general description	246
Venus SS2 monitor	229
description	231
Vertical	
deflection	267
sweep	223, 241
sync	237
sychronization	267
Video	
amplifier	150, 223
chain	216
circuits	241, 267
input selection	234
processing circuits	237

W

Waveforms, test point	142
Weather satellite	274
antenna	w93
operation at station	291
receiving station	276
WOLMD	
digital-scan converter	118
monitor	95
monitor, theory of operation	100
monitor, tuning	101
W9NTP slow-scan color converter	173

Z

Zoom lens	221